Student Edition

Eureka Math

Grade 3
Modules 1 & 2

Special thanks go to the Gordon A. Cain Center and to the Department of Mathematics at Louisiana State University for their support in the development of *Eureka Math*.

For a free *Eureka Math* Teacher Resource Pack, Parent Tip Sheets, and more please visit www.Eureka.tools

Published by the non-profit Great Minds

Copyright © 2015 Great Minds. No part of this work may be reproduced, sold, or commercialized, in whole or in part, without written permission from Great Minds. Non-commercial use is licensed pursuant to a Creative Commons Attribution-NonCommercial-ShareAlike 4.0 license; for more information, go to http://greatminds.net/maps/math/copyright. "Great Minds" and "Eureka Math" are registered trademarks of Great Minds.

Printed in the U.S.A.
This book may be purchased from the publisher at eureka-math.org
10 9 8 7 6

ISBN 978-1-63255-298-3

Name ______________________________ Date ______________

1. Fill in the blanks to make true statements.

a. 3 groups of five = _________

3 fives = _________

3×5 = _________

b. $3 + 3 + 3 + 3 + 3$ = _________

5 groups of three = _________

5×3 = _________

c. $6 + 6 + 6 + 6$ = __________

_______ groups of six = __________

$4 \times$ ______ = __________

d. $4 +$ ____ + ____ + ____ + ____ + ____ = _________

6 groups of ________ = ___________

$6 \times$ ______ = __________

©2015 Great Minds. eureka-math.org
G3-M1-SE-B1-1.3.1-01.2016

2. The picture below shows 2 groups of apples. Does the picture show 2×3? Explain why or why not.

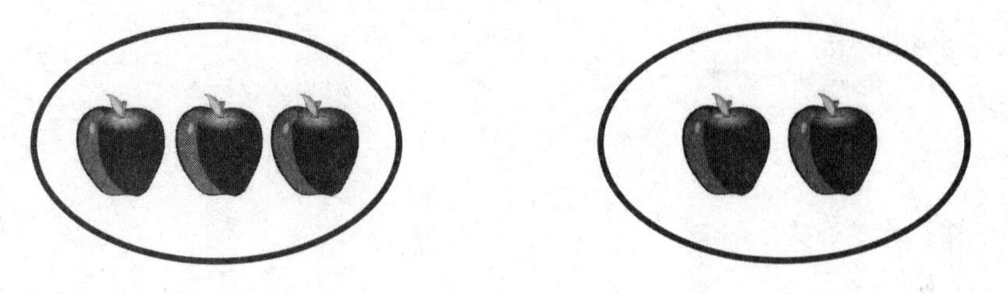

3. Draw a picture to show $2 \times 3 = 6$.

4. Caroline, Brian, and Marta share a box of chocolates. They each get the same amount. Circle the chocolates below to show 3 groups of 4. Then, write a repeated addition sentence and a multiplication sentence to represent the picture.

©2015 Great Minds. eureka-math.org
G3-M1-SE-B1-1.3.1-01.2016

Name ______________________________ Date ________________

1. Fill in the blanks to make true statements.

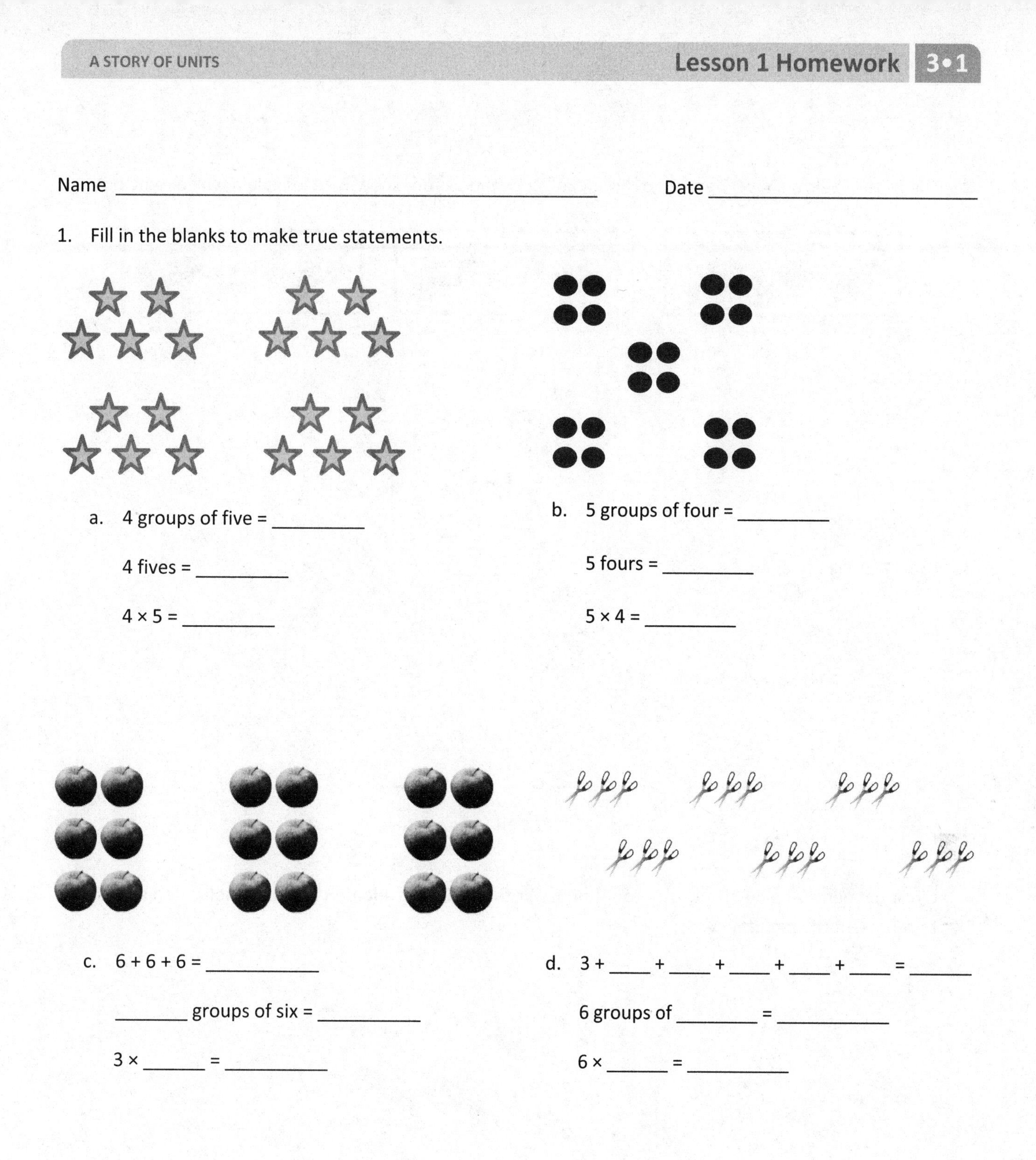

a. 4 groups of five = _________

4 fives = _________

4 × 5 = _________

b. 5 groups of four = _________

5 fours = _________

5 × 4 = _________

c. 6 + 6 + 6 = ___________

_______ groups of six = __________

3 × ______ = _________

d. 3 + ____ + ____ + ____ + ____ + ____ = ______

6 groups of ________ = ___________

6 × ______ = _________

©2015 Great Minds. eureka-math.org
G3-M1-SE-B1-1.3.1-01.2016

2. The picture below shows 3 groups of hot dogs. Does the picture show 3×3? Explain why or why not.

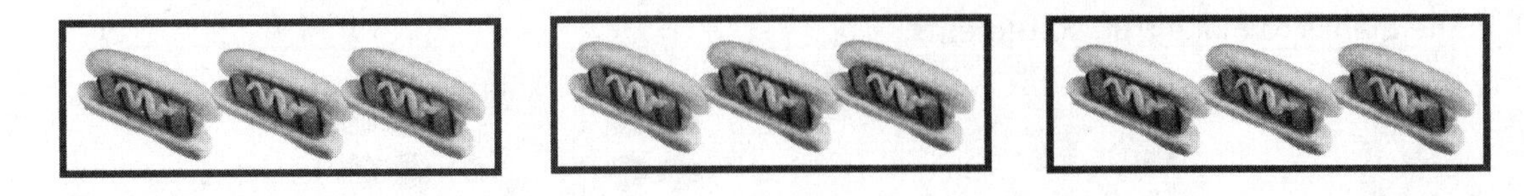

3. Draw a picture to show $4 \times 2 = 8$.

4. Circle the pencils below to show 3 groups of 6. Write a repeated addition and a multiplication sentence to represent the picture.

EUREKA MATH™
©2015 Great Minds. eureka-math.org
G3-M1-SE-B1-1.3.1-01.2016

Name ____________________________________ Date ____________________

Use the arrays below to answer each set of questions.

1. a. How many rows of cars are there? ________

   b. How many cars are there in each row? ________

2. a. What is the number of rows? ________

   b. What is the number of objects in each row? ________

3. a. There are 4 spoons in each row. How many spoons are in 2 rows? ________

   b. Write a multiplication expression to describe the array. ________________

4. a. There are 5 rows of triangles. How many triangles are in each row? ________

   b. Write a multiplication expression to describe the total number of triangles.

   ____

©2015 Great Minds. eureka-math.org
G3-M1-SE-B1-1.3.1-01.2016

5. The dots below show 2 groups of 5.

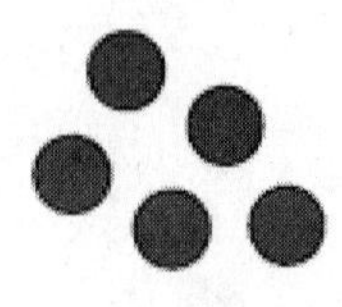

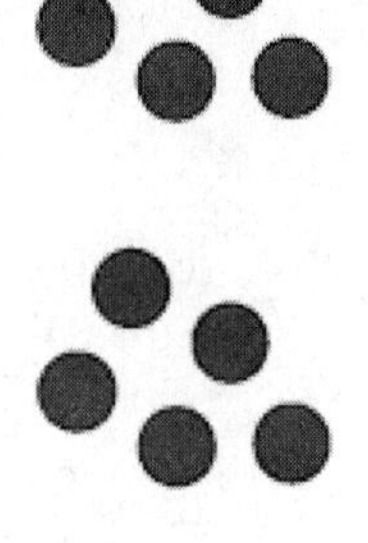

   a. Redraw the dots as an array that shows 2 rows of 5.

   b. Compare the drawing to your array. Write at least 1 reason why they are the same and 1 reason why they are different.

6. Emma collects rocks. She arranges them in 4 rows of 3. Draw Emma's array to show how many rocks she has altogether. Then, write a multiplication equation to describe the array.

7. Joshua organizes cans of food into an array. He thinks, "My cans show 5 × 3!" Draw Joshua's array to find the total number of cans he organizes.

©2015 Great Minds. eureka-math.org
G3-M1-SE-B1-1.3.1-01.2016

Name ______________________________ Date ______________

Use the arrays below to answer each set of questions.

1.  a. How many rows of erasers are there? __________

   b. How many erasers are there in each row? __________

2. 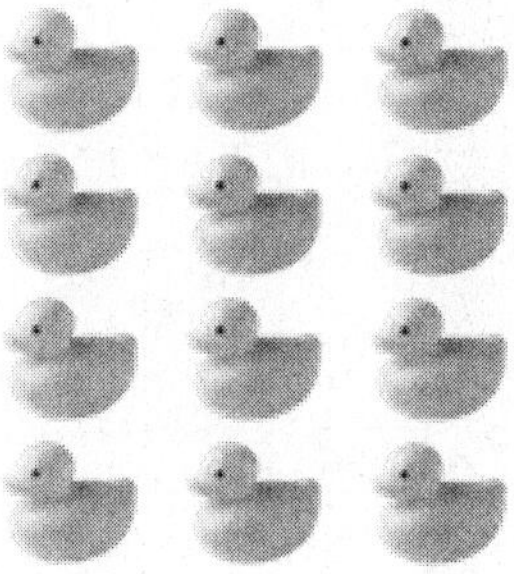 a. What is the number of rows? __________

   b. What is the number of objects in each row? __________

3. 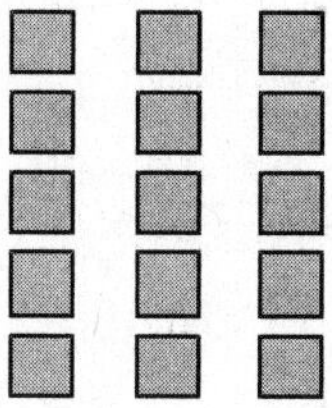 a. There are 3 squares in each row. How many squares are in 5 rows? __________

   b. Write a multiplication expression to describe the array. __________

4.  a. There are 6 rows of stars. How many stars are in each row? __________

   b. Write a multiplication expression to describe the array. __________

©2015 Great Minds. eureka-math.org
G3-M1-SE-B1-1.3.1-01.2016

5. The triangles below show 3 groups of four.

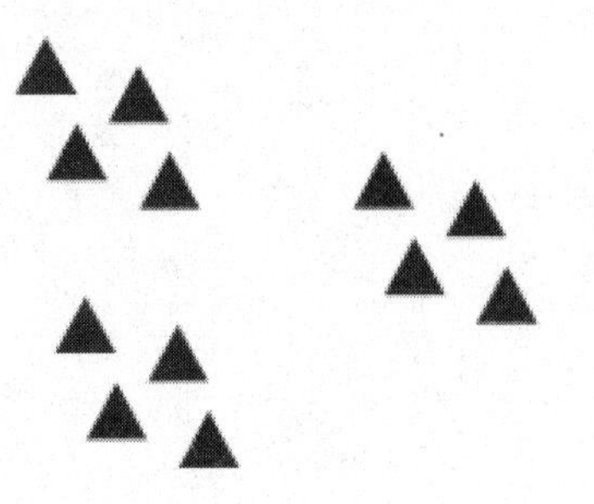

   a. Redraw the triangles as an array that shows 3 rows of four.

   b. Compare the drawing to your array. How are they the same? How are they different?

6. Roger has a collection of stamps. He arranges the stamps into 5 rows of four. Draw an array to represent Roger's stamps. Then, write a multiplication equation to describe the array.

7. Kimberly arranges her 18 markers as an array. Draw an array that Kimberly might make. Then, write a multiplication equation to describe your array.

©2015 Great Minds. eureka-math.org
G3-M1-SE-B1-1.3.1-01.2016

threes array

©2015 Great Minds. eureka-math.org
G3-M1-SE-B1-1.3.1-01.2016

Name ______________________________ Date ______________

Solve Problems 1–4 using the pictures provided for each problem.

1. There are 5 flowers in each bunch. How many flowers are in 4 bunches?

   a. Number of groups: __________ Size of each group: __________

   b. $4 \times 5 =$ __________

   c. There are __________ flowers altogether.

---

2. There are _______ candies in each box. How many candies are in 6 boxes?

   a. Number of groups: __________ Size of each group: __________

   b. $6 \times$ __________ $=$ __________

   c. There are __________ candies altogether.

---

3. There are 4 oranges in each row. How many oranges are there in ______ rows?

   a. Number of rows: __________ Size of each row: __________

   b. __________ $\times 4 =$ __________

   c. There are __________ oranges altogether.

©2015 Great Minds. eureka-math.org
G3-M1-SE-B1-1.3.1-01.2016

4. There are ________ loaves of bread in each row. How many loaves of bread are there in 5 rows?

a. Number of rows: ___________ Size of each row: __________

b. __________ × __________ = ___________

c. There are _________ loaves of bread altogether.

5. a. Write a multiplication equation for the array shown below.

X X X

X X X

X X X

X X X

b. Draw a number bond for the array where each part represents the amount in one row.

6. Draw an array using factors 2 and 3. Then, show a number bond where each part represents the amount in one row.

©2015 Great Minds. eureka-math.org
G3-M1-SE-B1-1.3.1-01.2016

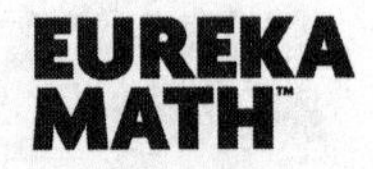

Name ______________________________ Date ________________

Solve Problems 1–4 using the pictures provided for each problem.

1. There are 5 pineapples in each group. How many pineapples are there in 5 groups?

a. Number of groups: ____________ Size of each group: ____________

b. $5 \times 5 =$ ____________

c. There are __________ pineapples altogether.

---

2. There are _______ apples in each basket. How many apples are there in 6 baskets?

a. Number of groups: ____________ Size of each group: ____________

b. $6 \times$ _________ = ____________

c. There are __________ apples altogether.

©2015 Great Minds. eureka-math.org
G3-M1-SE-B1-1.3.1-01.2016

3. There are 4 bananas in each row. How many bananas are there in ________ rows?

a. Number of rows: __________ Size of each row: __________

b. _________ × 4 = __________

c. There are __________ bananas altogether.

4. There are ___________ peppers in each row. How many peppers are there in 6 rows?

a. Number of rows: __________ Size of each row: __________

b. _________ × __________ = __________

c. There are __________ peppers altogether.

5. Draw an array using factors 4 and 2. Then, show a number bond where each part represents the amount in one row.

©2015 Great Minds. eureka-math.org
G3-M1-SE-B1-1.3.1-01.2016

Name ________________________________ Date ________________

1. 14 flowers are divided into 2 equal groups.

There are ________ flowers in each group.

2. 28 books are divided into 4 equal groups.

There are ________ books in each group.

3. 30 apples are divided into ______ equal groups.

There are ________ apples in each group.

4. _______ cups are divided into _______ equal groups.

There are ________ cups in each group.

$12 \div 2 =$ ________

5. There are ________ toys in each group.

$15 \div 3 =$ ________

6. $9 \div 3 =$ ________

©2015 Great Minds. eureka-math.org
G3-M1-SE-B1-1.3.1-01.2016

7. Audrina has 24 colored pencils. She puts them in 4 equal groups. How many colored pencils are in each group?

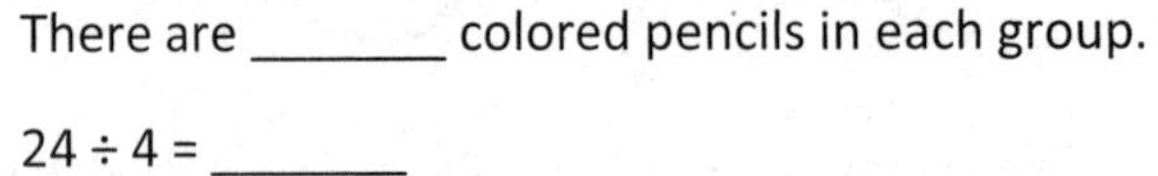

There are _______ colored pencils in each group.

24 ÷ 4 = _______

8. Charlie picks 20 apples. He divides them equally between 5 baskets. Draw the apples in each basket.

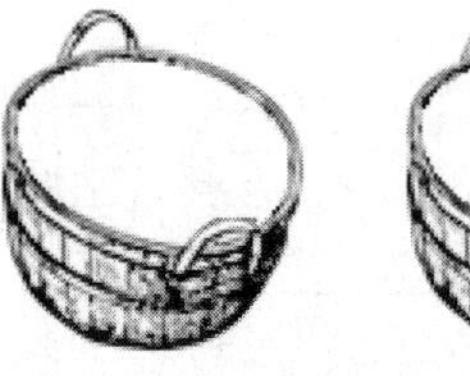 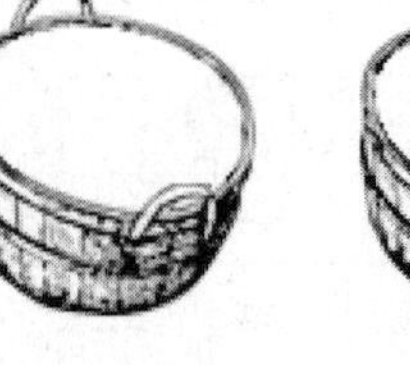 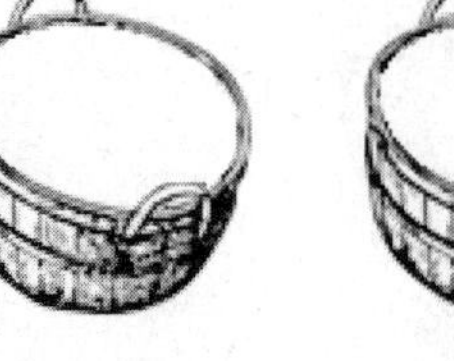 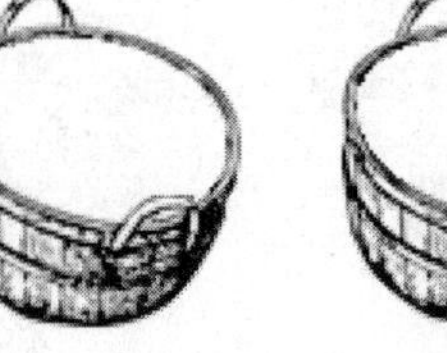 

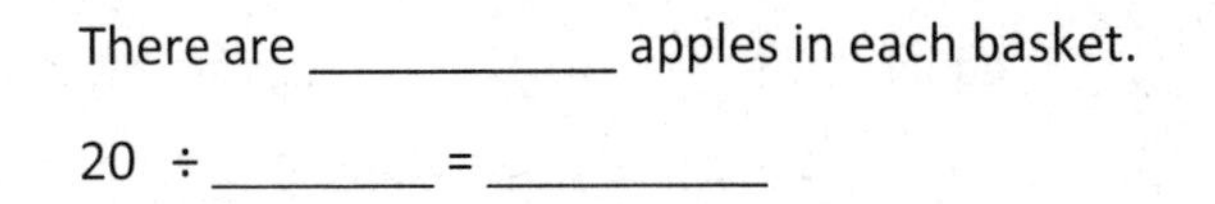

There are ___________ apples in each basket.

20 ÷ ________ = __________

9. Chelsea collects butterfly stickers. The picture shows how she placed them in her book. Write a division sentence to show how she equally grouped her stickers.

There are ____________ butterflies in each row.

__________ ÷ __________ = __________

©2015 Great Minds. eureka-math.org
G3-M1-SE-B1-1.3.1-01.2016

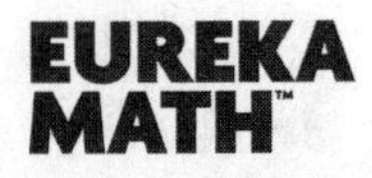

Name ______________________________ Date ____________________

1. 12 chairs are divided into 2 equal groups.

There are ________ chairs in each group.

2. 21 triangles are divided into 3 equal groups.

There are ________ triangles in each group.

3. 25 erasers are divided into ______ equal groups.

There are ________ erasers in each group.

4. _______ chickens are divided into _______ equal groups.

There are ________ chickens in each group.

$9 \div 3 =$ ________

5. There are ________ buckets in each group.

$12 \div 4 =$ ________

6. $16 \div 4 =$ ________

©2015 Great Minds. eureka-math.org
G3-M1-SE-B1-1.3.1-01.2016

7. Andrew has 21 keys. He puts them in 3 equal groups. How many keys are in each group?

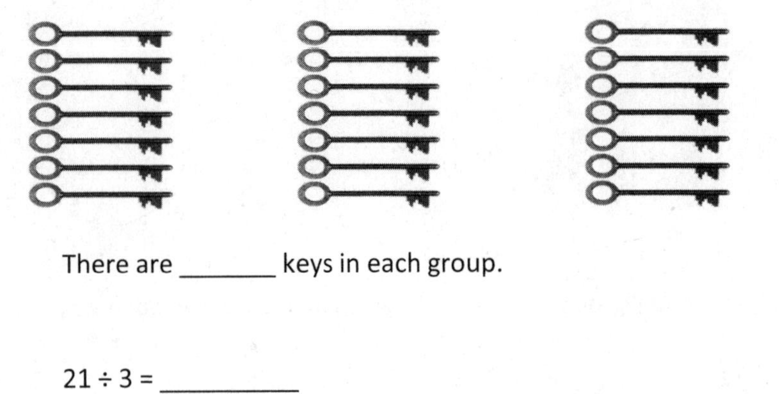

There are _______ keys in each group.

21 ÷ 3 = __________

8. Mr. Doyle has 20 pencils. He divides them equally between 4 tables. Draw the pencils on each table.

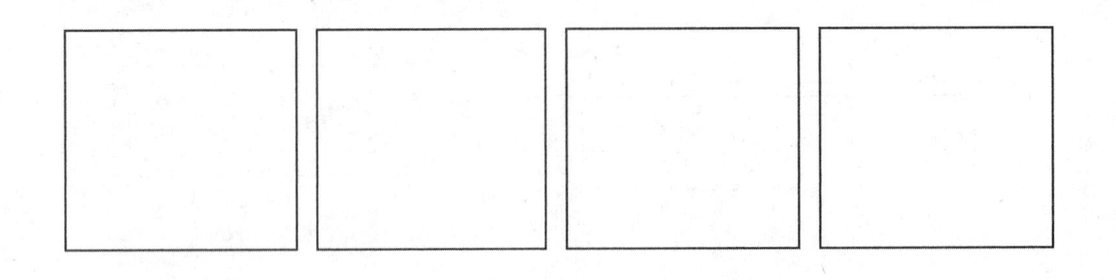

There are ___________ pencils on each table.

20 ÷ ________ = __________

9. Jenna has markers. The picture shows how she placed them on her desk. Write a division sentence to represent how she equally grouped her markers.

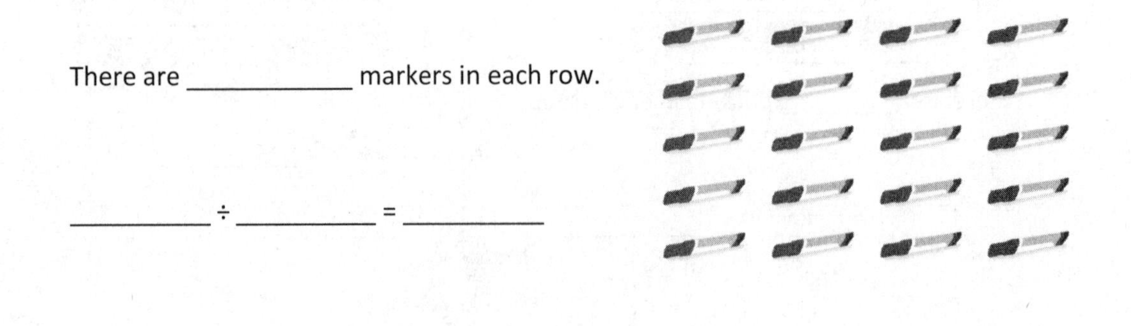

There are ____________ markers in each row.

__________ ÷ __________ = __________

©2015 Great Minds. eureka-math.org
G3-M1-SE-B1-1.3.1-01.2016

Name ______________________________ Date ______________

1.

Divide 6 tomatoes into groups of 3.

There are ________ groups of 3 tomatoes.

$6 \div 3 = 2$

2.

Divide 8 lollipops into groups of 2.

There are _______ groups.

$8 \div 2 =$ _______

3.

Divide 10 stars into groups of 5.

$10 \div 5 =$ _______

4.

Divide the shells to show $12 \div 3 =$ ________, where the unknown represents the number of groups.

How many groups are there? ________

©2015 Great Minds. eureka-math.org
G3-M1-SE-B1-1.3.1-01.2016

5. Rachel has 9 crackers. She puts 3 crackers in each bag. Circle the crackers to show Rachel's bags.

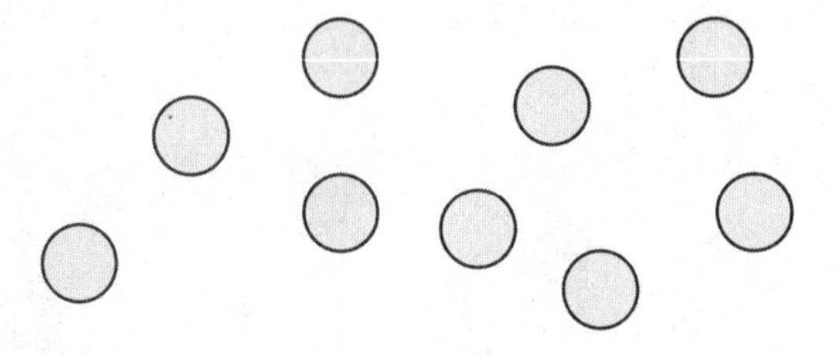

   a. Write a division sentence where the answer represents the number of Rachel's bags.

   b. Draw a number bond to represent the problem.

6. Jameisha has 16 wheels to make toy cars. She uses 4 wheels for each car.

   a. Use a count-by to find the number of cars Jameisha can build. Make a drawing to match your counting.

   b. Write a division sentence to represent the problem.

©2015 Great Minds. eureka-math.org
G3-M1-SE-B1-1.3.1-01.2016

Name ______________________________ Date ________________

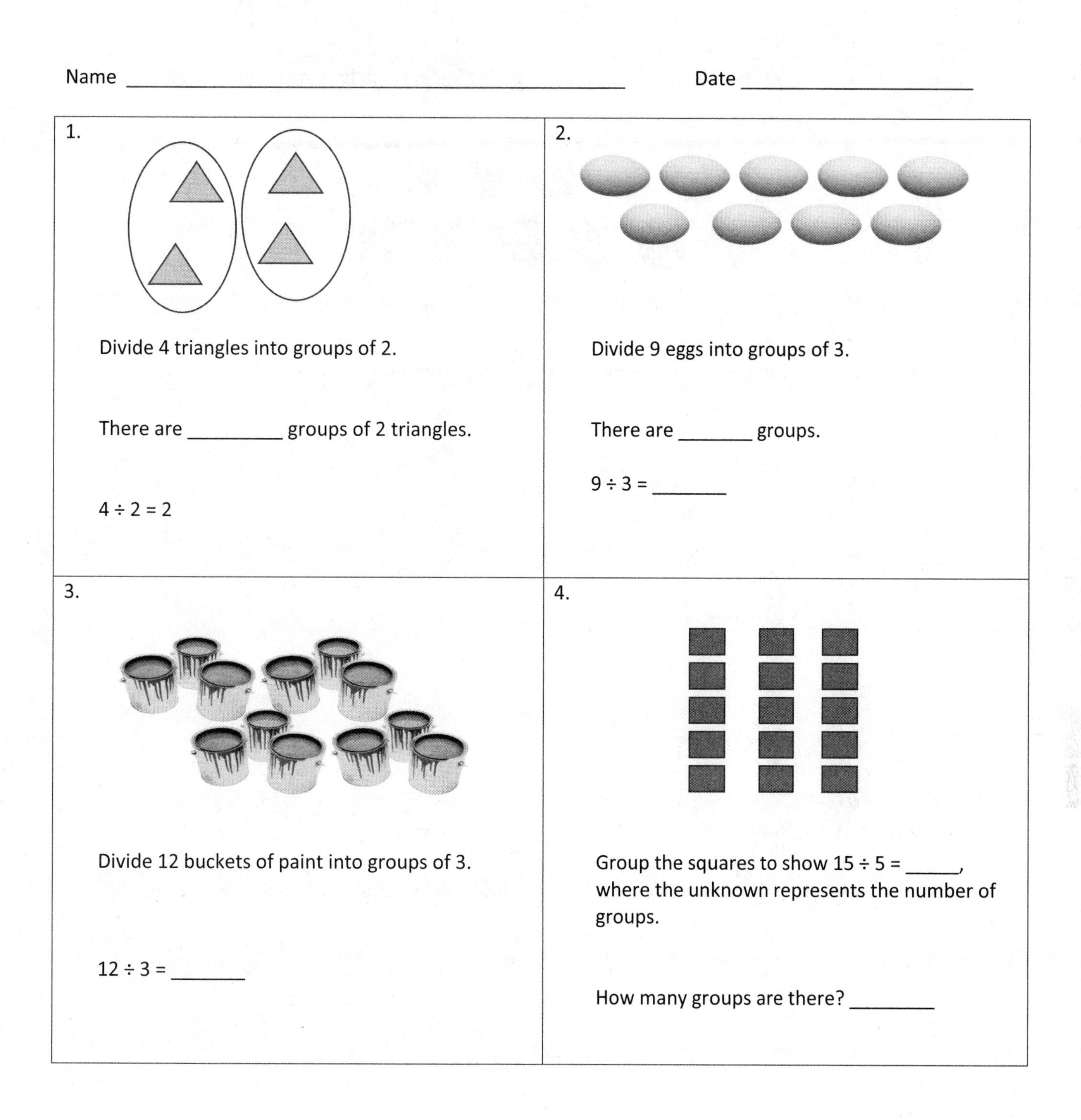

1.

Divide 4 triangles into groups of 2.

There are _________ groups of 2 triangles.

$4 \div 2 = 2$

2.

Divide 9 eggs into groups of 3.

There are _______ groups.

$9 \div 3 =$ _______

3.

Divide 12 buckets of paint into groups of 3.

$12 \div 3 =$ _______

4.

Group the squares to show $15 \div 5 =$ _____, where the unknown represents the number of groups.

How many groups are there? ________

©2015 Great Minds. eureka-math.org
G3-M1-SE-B1-1.3.1-01.2016

5. Daniel has 12 apples. He puts 6 apples in each bag. Circle the apples to find the number of bags Daniel makes.

   a. Write a division sentence where the answer represents the number of Daniel's bags.

   b. Draw a number bond to represent the problem.

6. Jacob draws cats. He draws 4 legs on each cat for a total of 24 legs.

   a. Use a count-by to find the number of cats Jacob draws. Make a drawing to match your counting.

   b. Write a division sentence to represent the problem.

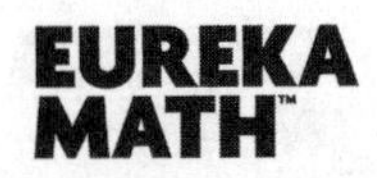

©2015 Great Minds. eureka-math.org
G3-M1-SE-B1-1.3.1-01.2016

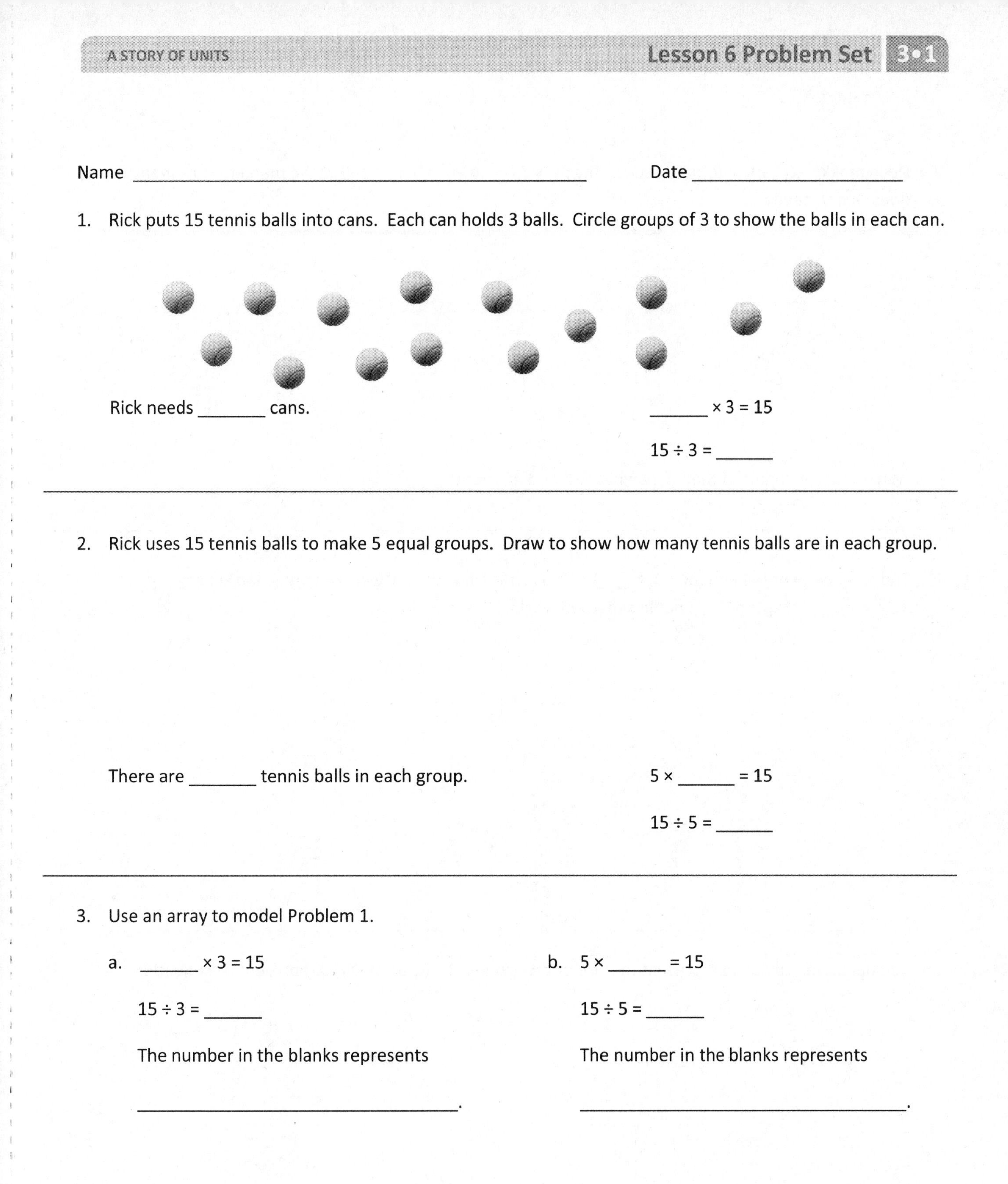

Name ________________________________ Date ________________

1. Rick puts 15 tennis balls into cans. Each can holds 3 balls. Circle groups of 3 to show the balls in each can.

Rick needs _______ cans.

______ × 3 = 15

15 ÷ 3 = ______

---

2. Rick uses 15 tennis balls to make 5 equal groups. Draw to show how many tennis balls are in each group.

There are _______ tennis balls in each group.

5 × ______ = 15

15 ÷ 5 = ______

---

3. Use an array to model Problem 1.

a. ______ × 3 = 15

15 ÷ 3 = ______

The number in the blanks represents

________________________________.

b. 5 × ______ = 15

15 ÷ 5 = ______

The number in the blanks represents

________________________________.

©2015 Great Minds. eureka-math.org
G3-M1-SE-B1-1.3.1-01.2016

4. Deena makes 21 jars of tomato sauce. She puts 7 jars in each box to sell at the market. How many boxes does Deena need?

   21 ÷ 7 = ______

   ______ × 7 = 21

   What is the meaning of the unknown factor and quotient? ____

   ____

5. The teacher gives the equation 4 × ____ = 12. Charlie finds the answer by writing and solving 12 ÷ 4 = ____. Explain why Charlie's method works.

6. The blanks in Problem 5 represent the size of the groups. Draw an array to represent the equations.

©2015 Great Minds. eureka-math.org
G3-M1-SE-B1-1.3.1-01.2016

Name ______________________________ Date ________________

1. Mr. Hannigan puts 12 pencils into boxes. Each box holds 4 pencils. Circle groups of 4 to show the pencils in each box.

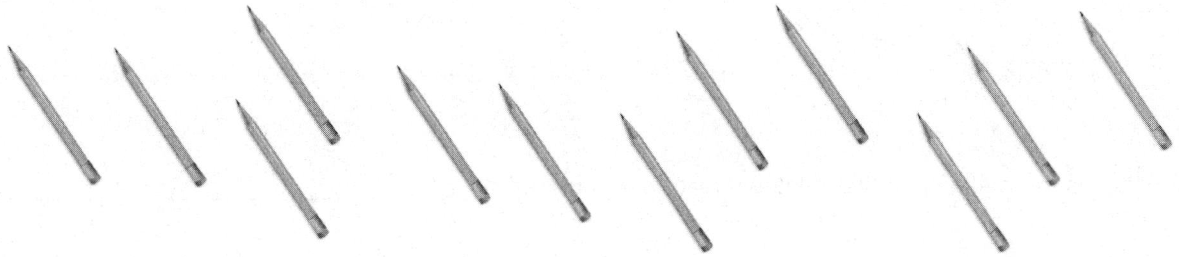

Mr. Hannigan needs _______ boxes.

______ × 4 = 12

12 ÷ 4 = ______

---

2. Mr. Hannigan places 12 pencils into 3 equal groups. Draw to show how many pencils are in each group.

There are _______ pencils in each group.

3 × ______ = 12

12 ÷ 3 = ______

---

3. Use an array to model Problem 1.

a. ______ × 4 = 12

12 ÷ 4 = ______

The number in the blanks represents

________________________________.

b. 3 × ______ = 12

12 ÷ 3 = ______

The number in the blanks represents

________________________________.

©2015 Great Minds. eureka-math.org
G3-M1-SE-B1-1.3.1-01.2016

4. Judy washes 24 dishes. She then dries and stacks the dishes equally into 4 piles. How many dishes are in each pile?

   $24 \div 4 = \_\_\_\_\_\_$

   $4 \times \_\_\_\_\_\_ = 24$

   What is the meaning of the unknown factor and quotient? ____________________________________

5. Nate solves the equation _____ × 5 = 15 by writing and solving 15 ÷ 5 = ____. Explain why Nate's method works.

6. The blanks in Problem 5 represent the number of groups. Draw an array to represent the equations.

©2015 Great Minds. eureka-math.org
G3-M1-SE-B1-1.3.1-01.2016

Name ______________________________ Date ______________

1. a. Draw an array that shows 6 rows of 2.

   b. Write a multiplication sentence where the first factor represents the number of rows.

   _______ × _______ = _______

2. a. Draw an array that shows 2 rows of 6.

   b. Write a multiplication sentence where the first factor represents the number of rows.

   _______ × _______ = _______

3. a. Turn your paper to look at the arrays in Problems 1 and 2 in different ways. What is the same and what is different about them?

   b. Why are the factors in your multiplication sentences in a different order?

4. Write a multiplication sentence for each expression. You might skip-count to find the totals.

   a. 6 twos: $6 \times 2 = 12$

   b. 2 sixes: ____________

   c. 7 twos: ____________

   d. 2 sevens: ____________

   e. 9 twos: ____________

   f. 2 nines: ____________

   Extension:

   g. 11 twos: ____________

   h. 2 twelves: ____________

EUREKA MATH™ Lesson 7: Demonstrate the commutativity of multiplication, and practice related facts by skip-counting objects in array models. 27

©2015 Great Minds. eureka-math.org
G3-M1-SE-B1-1.3.1-01.2016

5. Write and solve multiplication sentences where the second factor represents the size of the row.

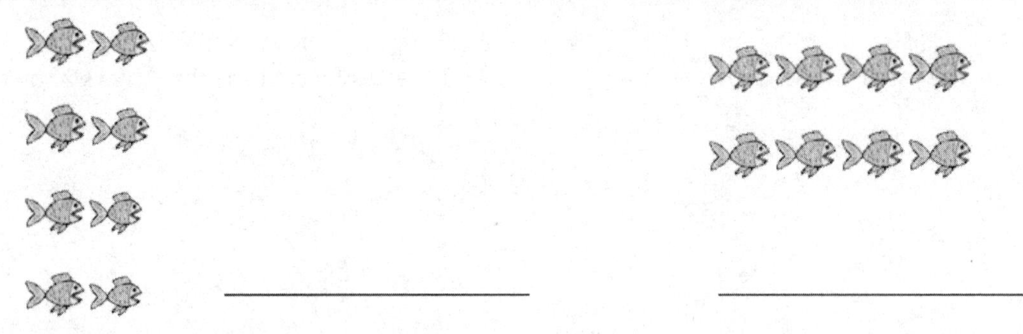

6. Ms. Nenadal writes $2 \times 7 = 7 \times 2$ on the board. Do you agree or disagree? Draw arrays to help explain your thinking.

7. Find the missing factor to make each equation true.

5 × 2 = 2 × _____ **_____ × 8 = 8 × 2** **2 × 10 = _____ × 2** **2 × _____ = 9 × 2**

8. Jada gets 2 new packs of erasers. Each pack has 6 erasers in it.
   a. Draw an array to show how many erasers Jada has altogether.

   b. Write and solve a multiplication sentence to describe the array.

   c. Use the commutative property to write and solve a different multiplication sentence for the array.

©2015 Great Minds. eureka-math.org
G3-M1-SE-B1-1.3.1-01.2016

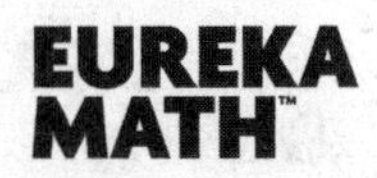

Name ______________________________ Date ______________

1. a. Draw an array that shows 7 rows of 2.

   b. Write a multiplication sentence where the first factor represents the number of rows.

   _______ × _______ = _______

2. a. Draw an array that shows 2 rows of 7.

   b. Write a multiplication sentence where the first factor represents the number of rows.

   _______ × _______ = _______

3. a. Turn your paper to look at the arrays in Problems 1 and 2 in different ways. What is the same and what is different about them?

   b. Why are the factors in your multiplication sentences in a different order?

4. Write a multiplication sentence to match the number of groups. Skip-count to find the totals. The first one is done for you.

   a. 2 twos: $2 \times 2 = 4$
   b. 3 twos: ____________
   c. 2 threes: ____________
   d. 2 fours: ____________
   e. 4 twos: ____________
   f. 5 twos: ____________
   g. 2 fives: ____________
   h. 6 twos: ____________
   i. 2 sixes: ____________

©2015 Great Minds. eureka-math.org
G3-M1-SE-B1-1.3.1-01.2016

5. Write and solve multiplication sentences where the second factor represents the size of the row.

_______________ _______________

6. Angel writes $2 \times 8 = 8 \times 2$ in his notebook. Do you agree or disagree? Draw arrays to help explain your thinking.

7. Find the missing factor to make each equation true.

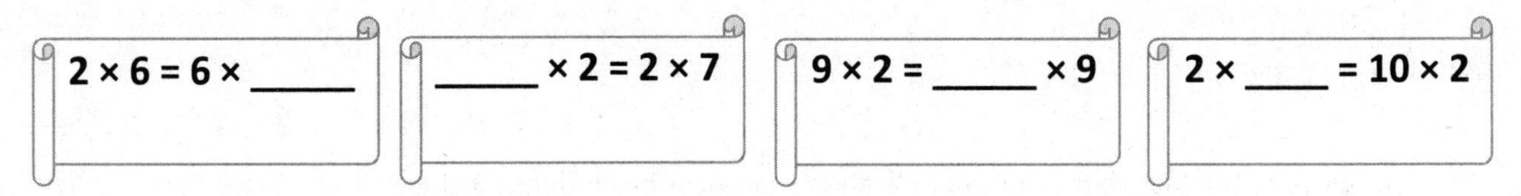

8. Tamia buys 2 bags of candy. Each bag has 7 pieces of candy in it.
   a. Draw an array to show how many pieces of candy Tamia has altogether.

   b. Write and solve a multiplication sentence to describe the array.

   c. Use the commutative property to write and solve a different multiplication sentence for the array.

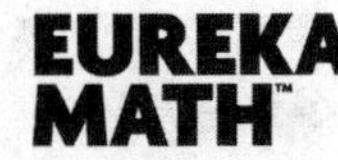

©2015 Great Minds. eureka-math.org
G3-M1-SE-B1-1.3.1-01.2016

Name ______________________________ Date ______________

1. Draw an array that shows 5 rows of 3.

2. Draw an array that shows 3 rows of 5.

3. Write multiplication expressions for the arrays in Problems 1 and 2. Let the first factor in each expression represent the number of rows. Use the commutative property to make sure the equation below is true.

_______ × _______ = _______ × _______

Problem 1 **Problem 2**

4. Write a multiplication sentence for each expression. You might skip-count to find the totals. The first one is done for you.

a. 2 threes: $2 \times 3 = 6$

d. 4 threes: ______________

g. 3 nines: ______________

b. 3 twos: ______________

e. 3 sevens: ______________

h. 9 threes: ______________

c. 3 fours: ______________

f. 7 threes: ______________

i. 10 threes: ______________

5. Find the unknowns that make the equations true. Then, draw a line to match related facts.

a. $3 + 3 + 3 + 3 + 3 =$ _______

d. $3 \times 8 =$ _______

b. $3 \times 9 =$ _______

e. _______ $= 5 \times 3$

c. 7 threes + 1 three = _______

f. $27 = 9 \times$ _______

©2015 Great Minds. eureka-math.org
G3-M1-SE-B1-1.3.1-01.2016

6. Isaac picks 3 tangerines from his tree every day for 7 days.

   a. Use circles to draw an array that represents the tangerines Isaac picks.

   b. How many tangerines does Isaac pick in 7 days? Write and solve a multiplication sentence to find the total.

   c. Isaac decides to pick 3 tangerines every day for 3 more days. Draw x's to show the new tangerines on the array in Part (a).

   d. Write and solve a multiplication sentence to find the total number of tangerines Isaac picks.

7. Sarah buys bottles of soap. Each bottle costs $2.

   a. How much money does Sarah spend if she buys 3 bottles of soap?

   ________ × ________ = $________

   b. How much money does Sarah spend if she buys 6 bottles of soap?

   ________ × ________ = $________

©2015 Great Minds. eureka-math.org
G3-M1-SE-B1-1.3.1-01.2016

Name ____________________ Date ____________

1. Draw an array that shows 6 rows of 3.

2. Draw an array that shows 3 rows of 6.

3. Write multiplication expressions for the arrays in Problems 1 and 2. Let the first factor in each expression represent the number of rows. Use the commutative property to make sure the equation below is true.

_______ × _______ = _______ × _______

Problem 1 **Problem 2**

4. Write a multiplication sentence for each expression. You might skip-count to find the totals. The first one is done for you.

a. 5 threes: $5 \times 3 = 15$

b. 3 fives: ____________

c. 6 threes: ____________

d. 3 sixes: ____________

e. 7 threes: ____________

f. 3 sevens: ____________

g. 8 threes: ____________

h. 3 nines: ____________

i. 10 threes: ____________

5. Find the unknowns that make the equations true. Then, draw a line to match related facts.

a. $3 + 3 + 3 + 3 + 3 + 3 =$ _______

b. $3 \times 5 =$ _______

c. 8 threes + 1 three = _______

d. $3 \times 9 =$ _______

e. _______ $= 6 \times 3$

f. $15 = 5 \times$ _______

©2015 Great Minds. eureka-math.org
G3-M1-SE-B1-1.3.1-01.2016

6. Fernando puts 3 pictures on each page of his photo album. He puts pictures on 8 pages.

   a. Use circles to draw an array that represents the total number of pictures in Fernando's photo album.

   b. Use your array to write and solve a multiplication sentence to find Fernando's total number of pictures.

   c. Fernando adds 2 more pages to his book. He puts 3 pictures on each new page. Draw x's to show the new pictures on the array in Part (a).

   d. Write and solve a multiplication sentence to find the new total number of pictures in Fernando's album.

7. Ivania recycles. She gets 3 cents for every can she recycles.

   a. How much money does Ivania make if she recycles 4 cans?

   __________ × __________ = ________ cents

   b. How much money does Ivania make if she recycles 7 cans?

   __________ × __________ = ________ cents

©2015 Great Minds. eureka-math.org
G3-M1-SE-B1-1.3.1-01.2016

Name ______________________________ Date ________________

1. The team organizes soccer balls into 2 rows of 5. The coach adds 3 rows of 5 soccer balls. Complete the equations to describe the total array.

a. $(5 + 5) + (5 + 5 + 5) =$ __________

b. 2 fives + ______ fives = __________ fives

c. ________ $\times 5 =$ __________

2. $7 \times 2 =$ ________

$5 \times 2 =$ ___

$2 \times 2 =$ ___

$10 + 4 =$ __________

__________ $\times 2 = 14$

3. $9 \times 2 =$ ________

$10 \times 2 =$ ___

$1 \times 2 =$ ___

$20 -$ __________ $= 18$

$9 \times 2 =$ __________

©2015 Great Minds. eureka-math.org
G3-M1-SE-B1-1.3.1-01.2016

4. Matthew organizes his baseball cards in 4 rows of 3.

   a. Draw an array that represents Matthew's cards using an x to show each card.

   b. Solve the equation to find Matthew's total number of cards. $4 \times 3 =$ ______

5. Matthew adds 2 more rows. Use circles to show his new cards on the array in Problem 4(a).

   a. Write and solve a multiplication equation to represent the circles you added to the array.

   ______ $\times 3 =$ ______

   b. Add the totals from the equations in Problems 4(b) and 5(a) to find Matthew's total cards.

   ______ + ______ $= 18$

   c. Write the multiplication equation that shows Matthew's total number of cards.

   ______ $\times$ ______ $= 18$

©2015 Great Minds. eureka-math.org
G3-M1-SE-B1-1.3.1-01.2016

Name ____________________ Date ____________

1. Dan organizes his stickers into 3 rows of four. Irene adds 2 more rows of stickers. Complete the equations to describe the total number of stickers in the array.

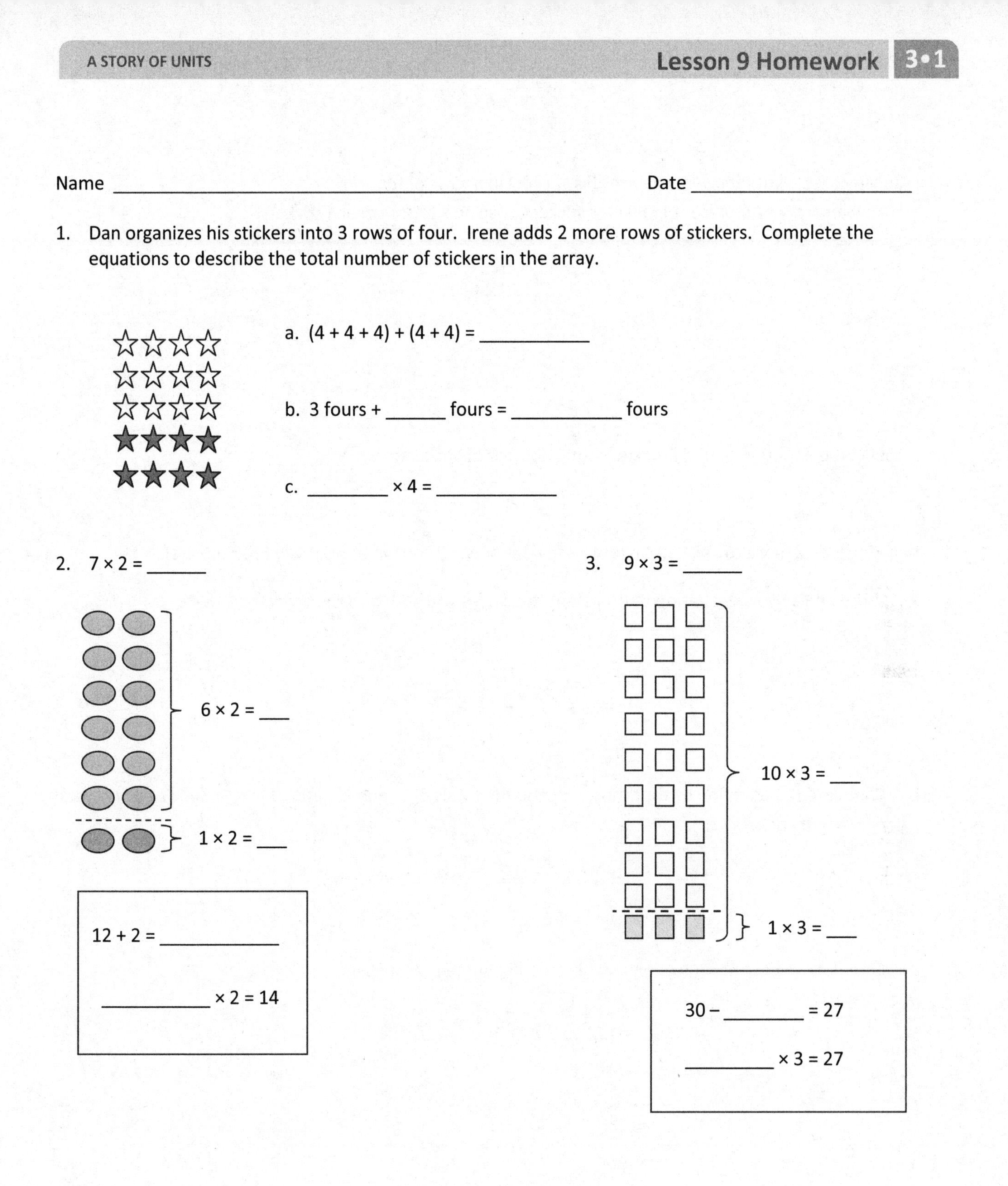

a. (4 + 4 + 4) + (4 + 4) = ________

b. 3 fours + _____ fours = ________ fours

c. ______ × 4 = ________

2. 7 × 2 = _____

6 × 2 = ___

1 × 2 = ___

12 + 2 = ________

________ × 2 = 14

3. 9 × 3 = _____

10 × 3 = ___

1 × 3 = ___

30 – ______ = 27

______ × 3 = 27

©2015 Great Minds. eureka-math.org
G3-M1-SE-B1-1.3.1-01.2016

4. Franklin collects stickers. He organizes his stickers in 5 rows of four.
   a. Draw an array to represent Franklin's stickers. Use an x to show each sticker.

   b. Solve the equation to find Franklin's total number of stickers. $5 \times 4 =$ ______

5. Franklin adds 2 more rows. Use circles to show his new stickers on the array in Problem 4(a).

   a. Write and solve an equation to represent the circles you added to the array.

   ______ $\times 4 =$ ______

   b. Complete the equation to show how you add the totals of 2 multiplication facts to find Franklin's total number of stickers.

   ______ + ______ = 28

   c. Complete the unknown to show Franklin's total number of stickers.

   ______ $\times 4 = 28$

©2015 Great Minds. eureka-math.org
G3-M1-SE-B1-1.3.1-01.2016

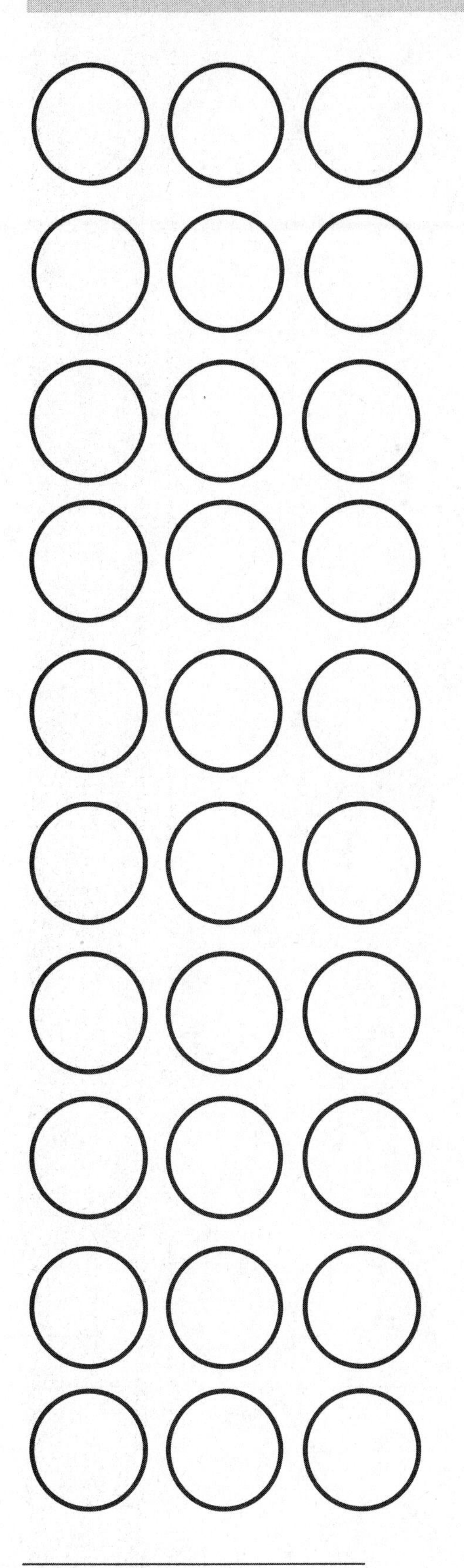

threes array no fill

©2015 Great Minds. eureka-math.org
G3-M1-SE-B1-1.3.1-01.2016

Name ____________________ Date ____________

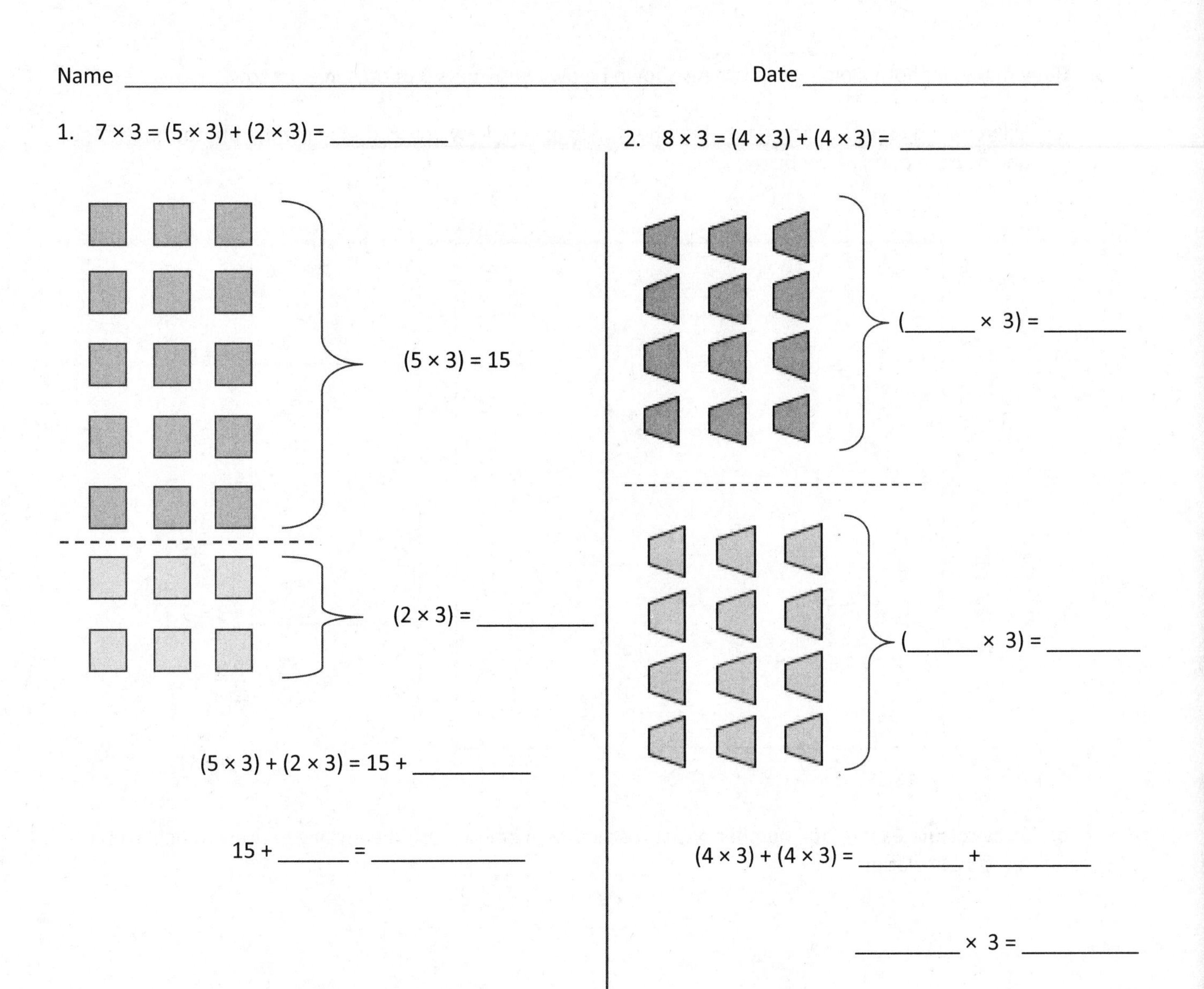

1. $7 \times 3 = (5 \times 3) + (2 \times 3) =$ ________

$(5 \times 3) = 15$

$(2 \times 3) =$ ________

$(5 \times 3) + (2 \times 3) = 15 +$ ________

$15 +$ ______ $=$ ________

2. $8 \times 3 = (4 \times 3) + (4 \times 3) =$ ______

(______ $\times 3) =$ ______

(______ $\times 3) =$ ______

$(4 \times 3) + (4 \times 3) =$ ________ $+$ ________

________ $\times 3 =$ ________

Lesson 10: Model the distributive property with arrays to decompose units as a strategy to multiply.

41

©2015 Great Minds. eureka-math.org
G3-M1-SE-B1-1.3.1-01.2016

3. Ruby makes a photo album. One page is shown below. Ruby puts 3 photos in each row.

   a. Fill in the equations on the right. Use them to help you draw arrays that show the photos on the top and bottom parts of the page.

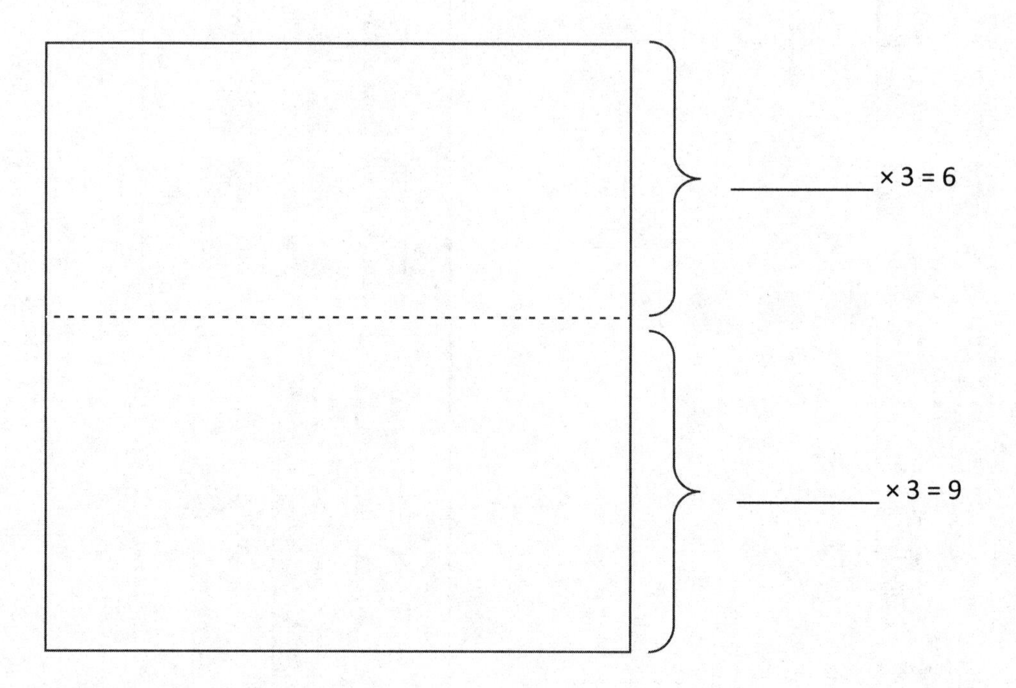

   b. Ruby calculates the total number of photos as shown below. Use the array you drew to help explain Ruby's calculation.

5 × 3 = 6 + 9 = 15

©2015 Great Minds. eureka-math.org
G3-M1-SE-B1-1.3.1-01.2016

Name ______________________________ Date ______________

1. 6 × 3 = ________

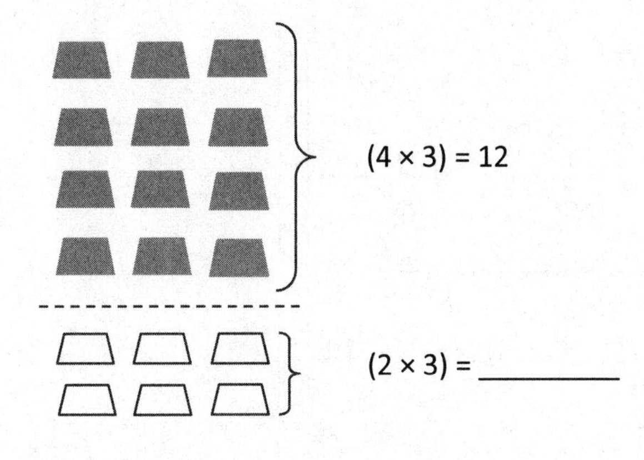

12 + ______ = ______

6 × 3 = ______

2. 8 × 2 = ________

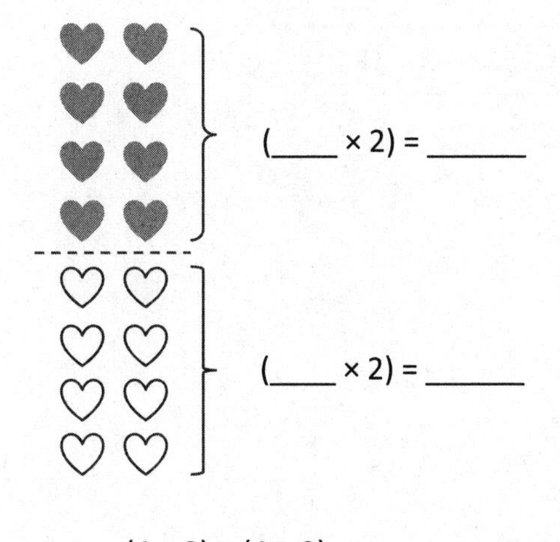

(4 × 2) + (4 × 2) = ______ + ______

____ × 2 = ________

Lesson 10: Model the distributive property with arrays to decompose units as a strategy to multiply.

43

©2015 Great Minds. eureka-math.org
G3-M1-SE-B1-1.3.1-01.2016

3. Adriana organizes her books on shelves. She puts 3 books in each row.

a. Fill in the equations on the right. Use them to draw arrays that show the books on Adriana's top and bottom shelves.

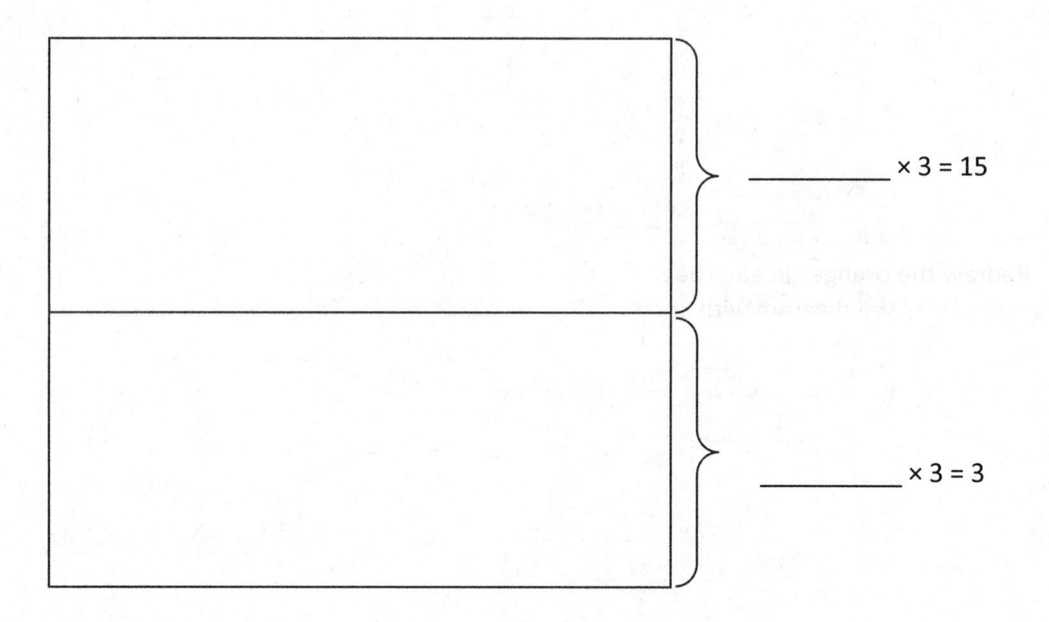

b. Adriana calculates the total number of books as shown below. Use the array you drew to help explain Adriana's calculation.

6 × 3 = 15 + 3 = 18

©2015 Great Minds. eureka-
G3-M1-SE-B1-1.3.1-01.2016

Name ______________________________ Date ________________

1. Mrs. Prescott has 12 oranges. She puts 2 oranges in each bag. How many bags does she have?

   a. Draw an array where each column shows a bag of oranges.

   _______ ÷ 2 = _______

   b. Redraw the oranges in each bag as a unit in the tape diagram. The first unit is done for you. As you draw, label the diagram with known and unknown information from the problem.

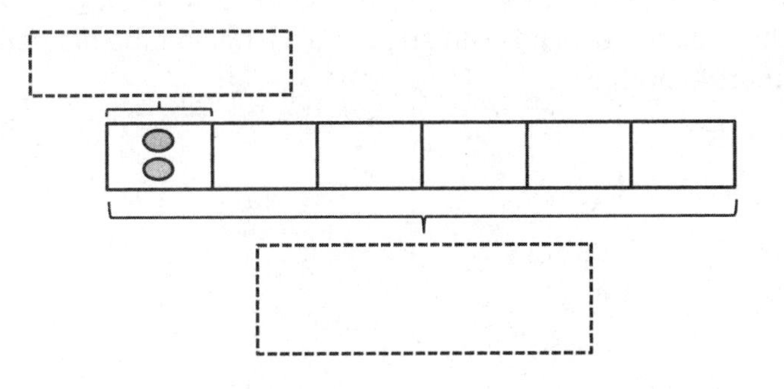

2. Mrs. Prescott arranges 18 plums into 6 bags. How many plums are in each bag? Model the problem with both an array and a labeled tape diagram. Show each column as the number of plums in each bag.

   There are _______ plums in each bag.

©2015 Great Minds. eureka-math.org
G3-M1-SE-B1-1.3.1-01.2016

3. Fourteen shopping baskets are stacked equally in 7 piles. How many baskets are in each pile? Model the problem with both an array and a labeled tape diagram. Show each column as the number of baskets in each pile.

4. In the back of the store, Mr. Prescott packs 24 bell peppers equally into 8 bags. How many bell peppers are in each bag? Model the problem with both an array and a labeled tape diagram. Show each column as the number of bell peppers in each bag.

5. Olga saves $2 a week to buy a toy car. The car costs $16. How many weeks will it take her to save enough to buy the toy?

©2015 Great Minds. eureka-math.org
G3-M1-SE-B1-1.3.1-01.2016

Name ______________________________ Date ________________

1. Fred has 10 pears. He puts 2 pears in each basket. How many baskets does he have?

   a. Draw an array where each column represents the number of pears in each basket.

   _______ ÷ 2 = _______

   b. Redraw the pears in each basket as a unit in the tape diagram. Label the diagram with known and unknown information from the problem.

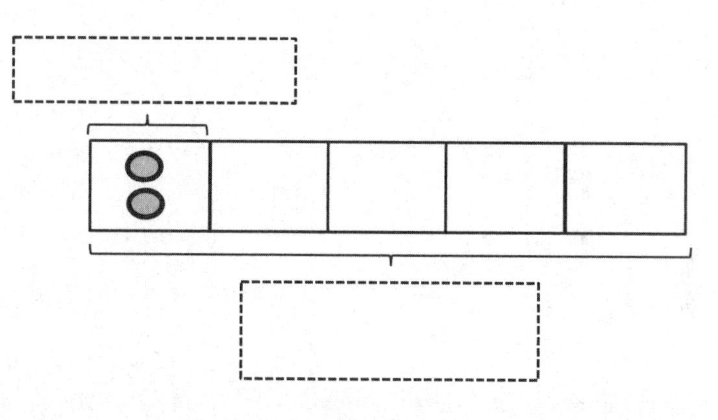

2. Ms. Meyer organizes 15 clipboards equally into 3 boxes. How many clipboards are in each box? Model the problem with both an array and a labeled tape diagram. Show each column as the number of clipboards in each box.

There are ________ clipboards in each box.

©2015 Great Minds. eureka-math.org
G3-M1-SE-B1-1.3.1-01.2016

3. Sixteen action figures are arranged equally on 2 shelves. How many action figures are on each shelf? Model the problem with both an array and a labeled tape diagram. Show each column as the number of action figures on each shelf.

4. Jasmine puts 18 hats away. She puts an equal number of hats on 3 shelves. How many hats are on each shelf? Model the problem with both an array and a labeled tape diagram. Show each column as the number of hats on each shelf.

5. Corey checks out 2 books a week from the library. How many weeks will it take him to check out a total of 14 books?

©2015 Great Minds. eureka-math.org
G3-M1-SE-B1-1.3.1-01.2016

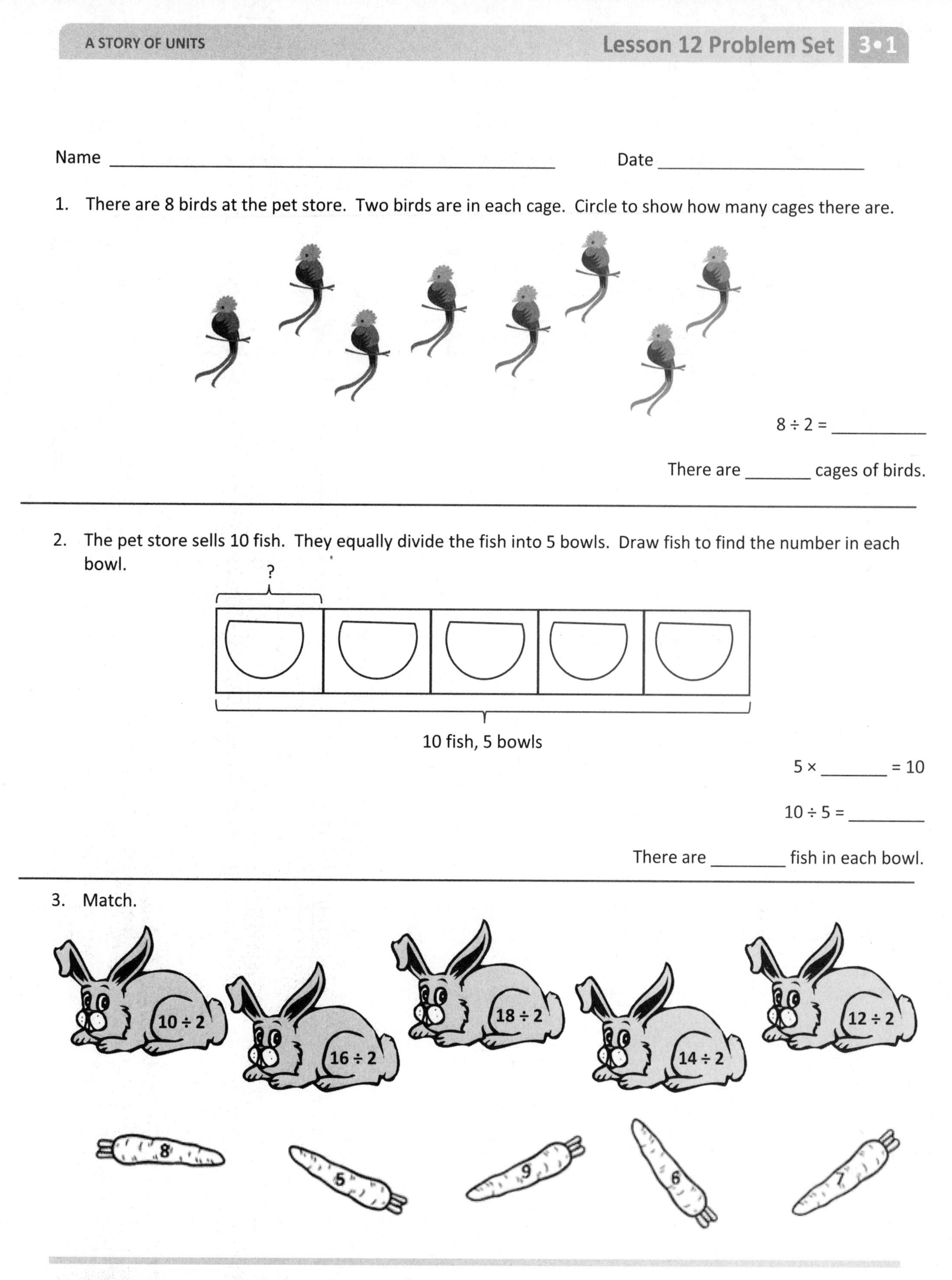

Name ______________________________ Date ____________________

1. There are 8 birds at the pet store. Two birds are in each cage. Circle to show how many cages there are.

$8 \div 2 =$ __________

There are _______ cages of birds.

---

2. The pet store sells 10 fish. They equally divide the fish into 5 bowls. Draw fish to find the number in each bowl.

?

10 fish, 5 bowls

$5 \times$ _______ $= 10$

$10 \div 5 =$ ________

There are ________ fish in each bowl.

---

3. Match.

10 ÷ 2
16 ÷ 2
18 ÷ 2
14 ÷ 2
12 ÷ 2

8
5
9
6
7

©2015 Great Minds. eureka-math.org
G3-M1-SE-B1-1.3.1-01.2016

4. Laina buys 14 meters of ribbon. She cuts her ribbon into 2 equal pieces. How many meters long is each piece? Label the tape diagram to represent the problem, including the unknown.

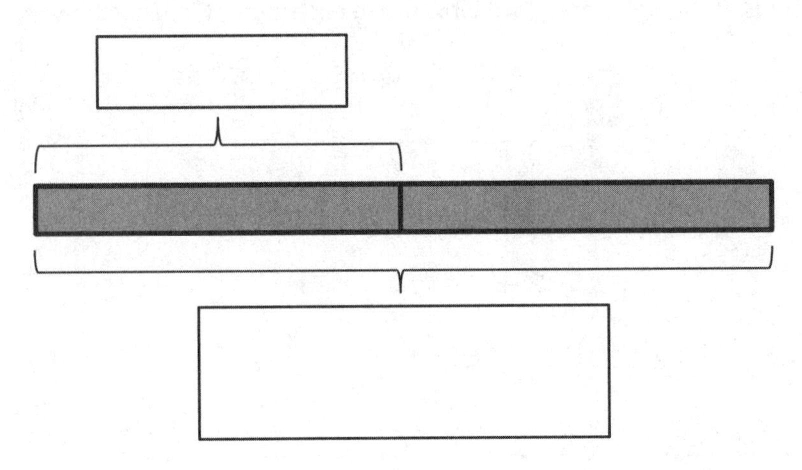

Each piece is __________ meters long.

5. Roy eats 2 cereal bars every morning. Each box has a total of 12 bars. How many days will it take Roy to finish 1 box?

6. Sarah and Esther equally share the cost of a present. The present costs $18. How much does Sarah pay?

©2015 Great Minds. eureka-math.org
G3-M1-SE-B1-1.3.1-01.2016

Name ______________________________ Date ______________

1. Ten people wait in line for the roller coaster. Two people sit in each car. Circle to find the total number of cars needed.

$10 \div 2 =$ ________

There are ______ cars needed.

---

2. Mr. Ramirez divides 12 frogs equally into 6 groups for students to study. Draw frogs to find the number in each group. Label known and unknown information on the tape diagram to help you solve.

$6 \times$ ______ $= 12$

$12 \div 6 =$ ______

There are ______ frogs in each group.

---

3. Match.

©2015 Great Minds. eureka-math.org
G3-M1-SE-B1-1.3.1-01.2016

4. Betsy pours 16 cups of water to equally fill 2 bottles. How many cups of water are in each bottle? Label the tape diagram to represent the problem, including the unknown.

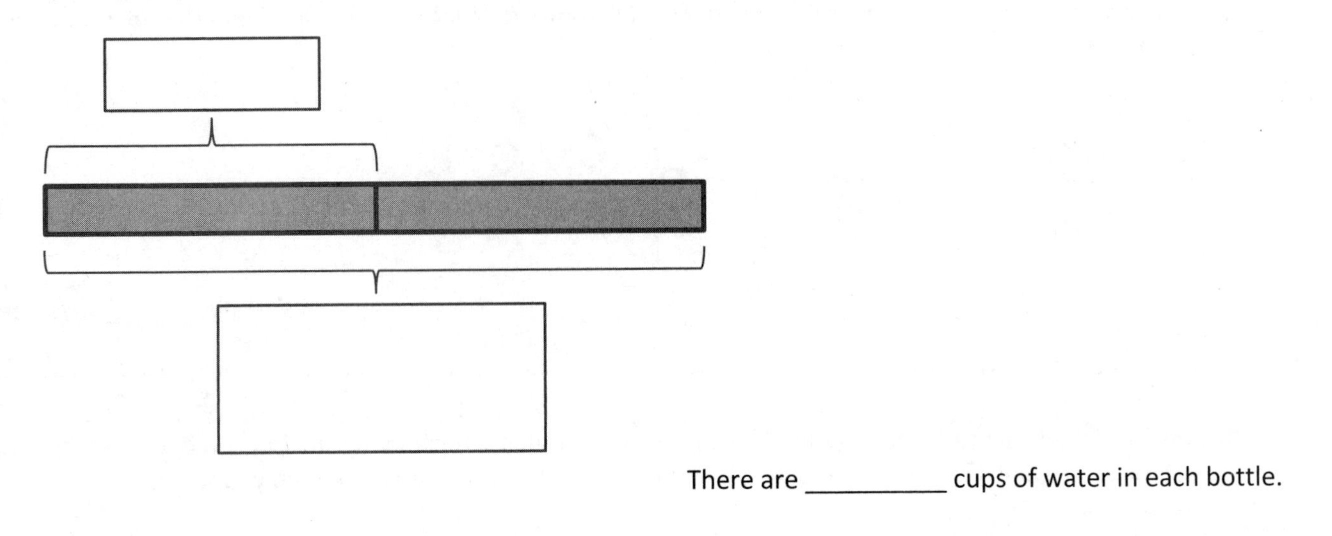

There are ________ cups of water in each bottle.

5. An earthworm tunnels 2 centimeters into the ground each day. The earthworm tunnels at about the same pace every day. How many days will it take the earthworm to tunnel 14 centimeters?

6. Sebastian and Teshawn go to the movies. The tickets cost $16 in total. The boys share the cost equally. How much does Teshawn pay?

©2015 Great Minds. eureka-math.org
G3-M1-SE-B1-1.3.1-01.2016

Name ______________________________ Date ________________

1. Fill in the blanks to make true number sentences.

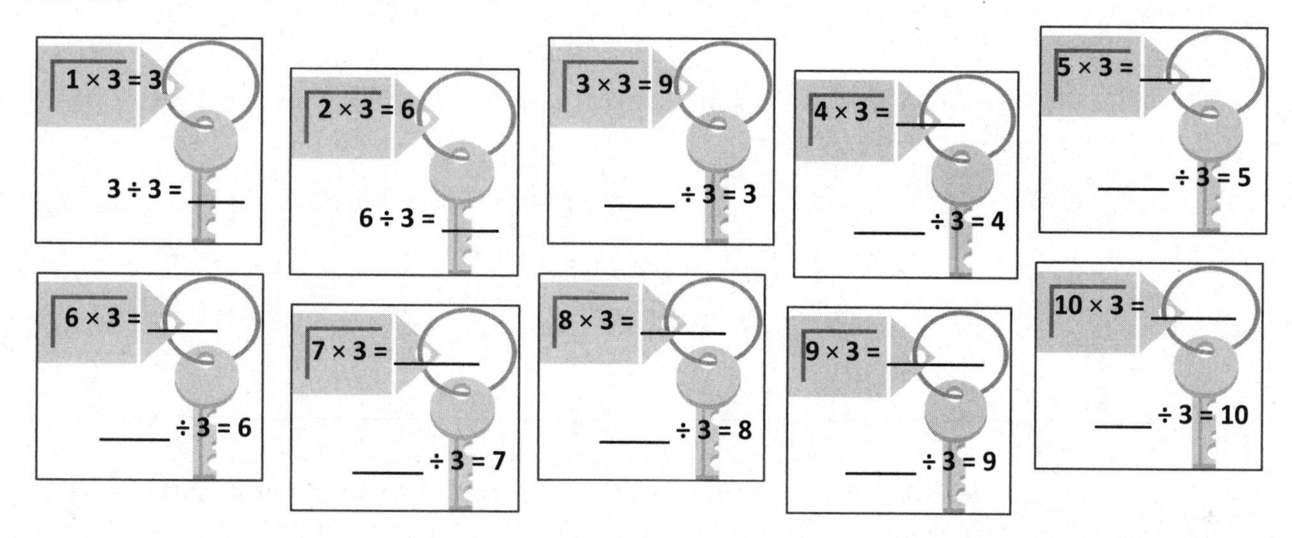

2. Mr. Lawton picks tomatoes from his garden. He divides the tomatoes into bags of 3.

   a. Circle to show how many bags he packs. Then, skip-count to show the total number of tomatoes.

   b. Draw and label a tape diagram to represent the problem.

____________ ÷ 3 = ____________

Mr. Lawton packs _______ bags of tomatoes.

©2015 Great Minds. eureka-math.org
G3-M1-SE-B1-1.3.1-01.2016

3. Camille buys a sheet of stamps that measures 15 centimeters long. Each stamp is 3 centimeters long. How many stamps does Camille buy? Draw and label a tape diagram to solve.

Camille buys _________ stamps.

---

4. Thirty third-graders go on a field trip. They are equally divided into 3 vans. How many students are in each van?

---

5. Some friends spend $24 altogether on frozen yogurt. Each person pays $3. How many people buy frozen yogurt?

©2015 Great Minds. eureka-math.org
G3-M1-SE-B1-1.3.1-01.2016

Name ______________________________ Date ____________________

1. Skip-count by fours. Match each answer to the appropriate expression.

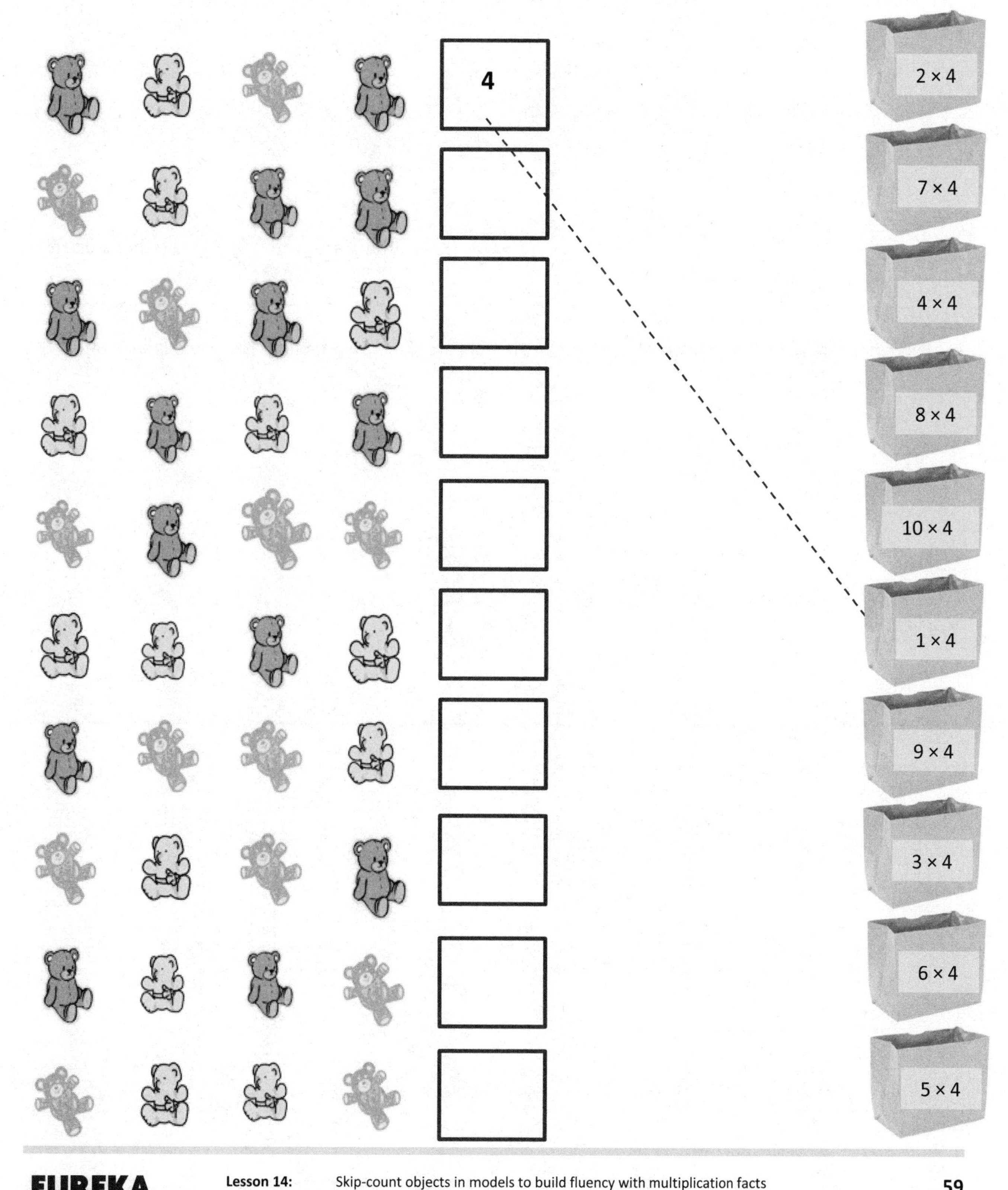

©2015 Great Minds. eureka-math.org
G3-M1-SE-B1-1.3.1-01.2016

2. Lisa places 5 rows of 4 juice boxes in the refrigerator. Draw an array and skip-count to find the total number of juice boxes.

There are __________ juice boxes in total.

---

3. Six folders are placed on each table. How many folders are there on 4 tables? Draw and label a tape diagram to solve.

---

4. Find the total number of corners on 8 squares.

©2015 Great Minds. eureka-math.org
G3-M1-SE-B1-1.3.1-01.2016

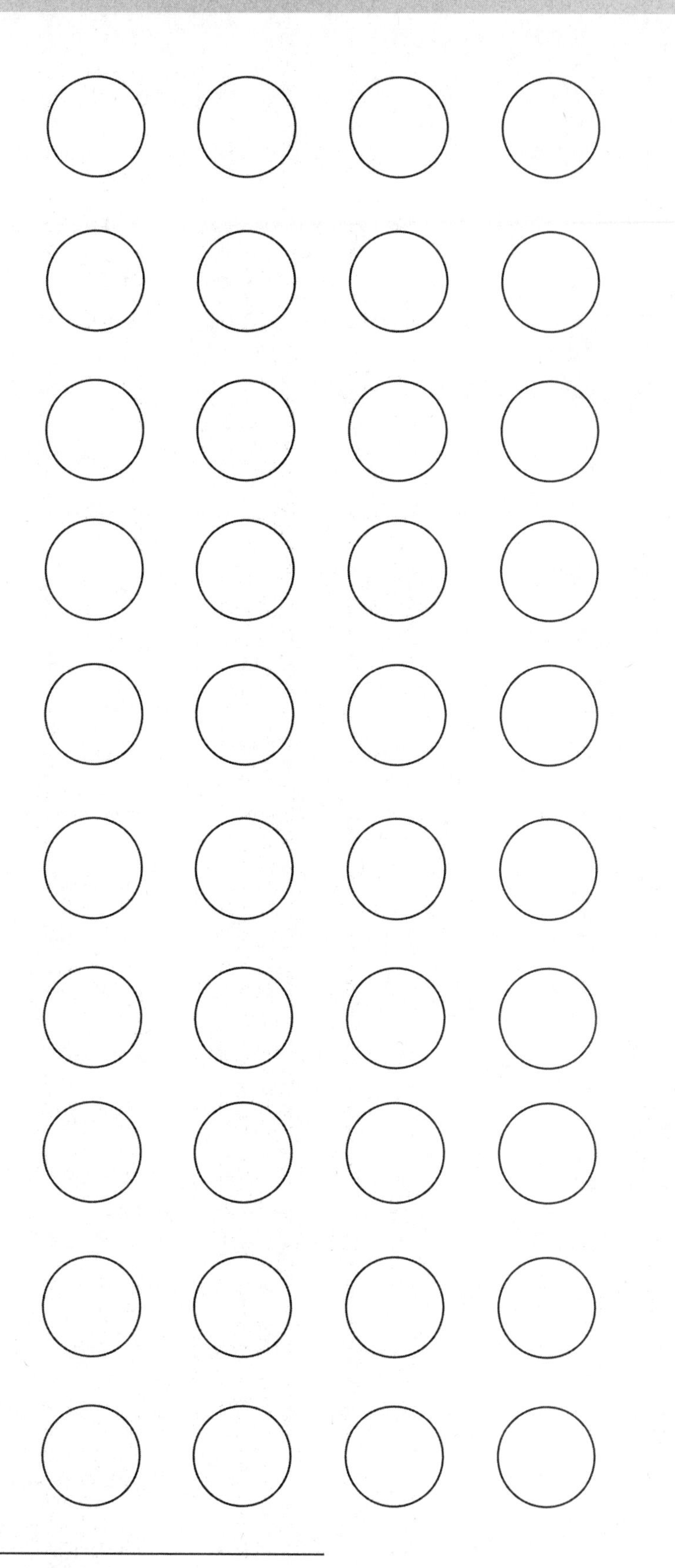

fours array

©2015 Great Minds. eureka-math.org
G3-M1-SE-B1-1.3.1-01.2016

Name ______________________________ Date ________________

1. Label the tape diagrams and complete the equations. Then, draw an array to represent the problems.

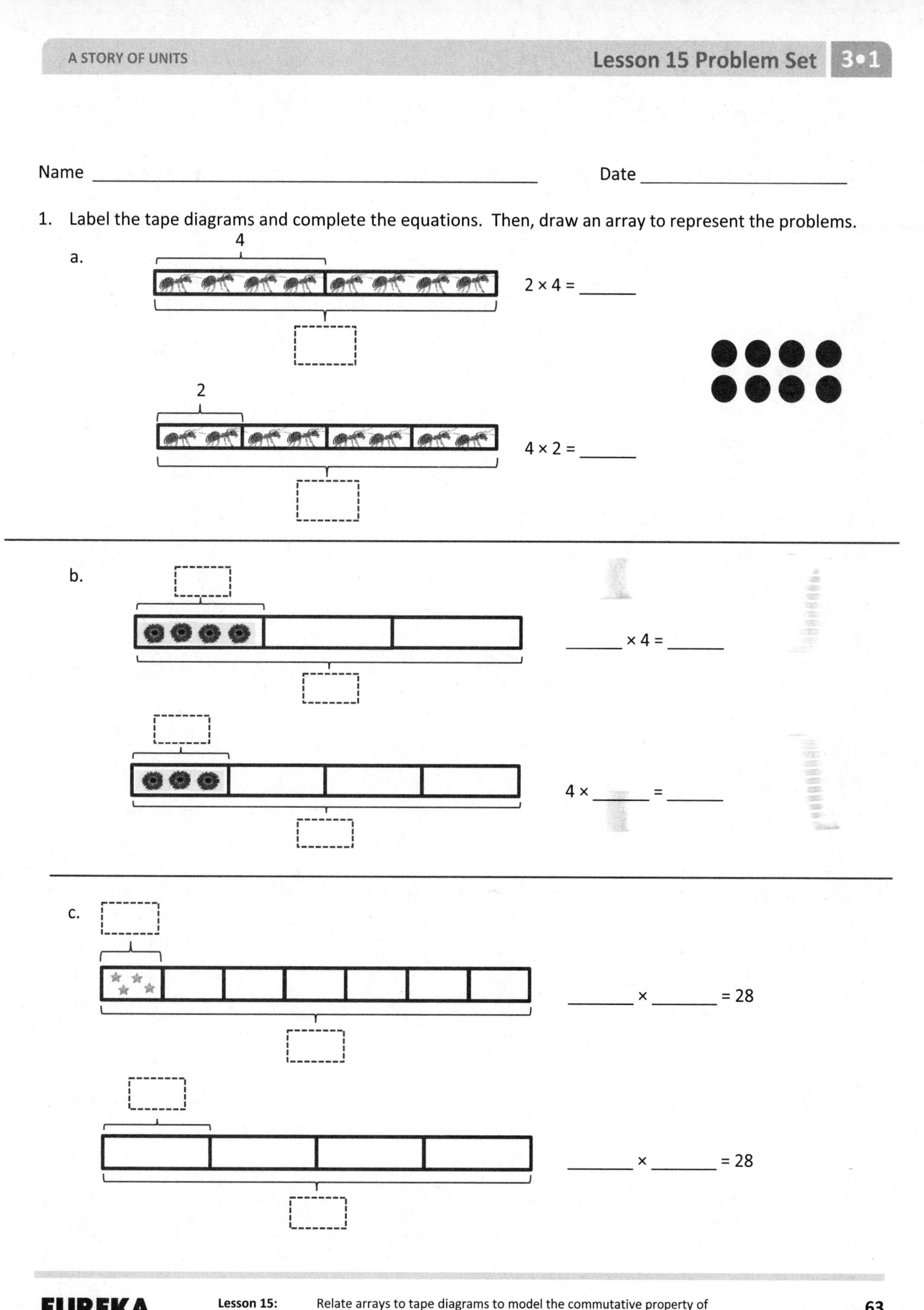

a.

4

$2 \times 4 =$ _____

©2015 Great Minds. eureka-math.org

$4 \times 2 =$ _____

b.

_____ $\times 4 =$ _____

$4 \times$ _____ $=$ _____

c.

_____ \times _____ $= 28$

_____ \times _____ $= 28$

EUREKA
MATH™

Lesson 15: Relate arrays to tape diagrams to model the commutative property of multiplication.

G3-M1-SE-B1-1.3.1-01.2016

2. Draw and label 2 tape diagrams to model why the statement in the box is true.

$4 \times 6 = 6 \times 4$

---

3. Grace picks 4 flowers from her garden. Each flower has 8 petals. Draw and label a tape diagram to show how many petals there are in total.

---

4. Michael counts 8 chairs in his dining room. Each chair has 4 legs. How many chair legs are there altogether?

©2015 Great Minds. eureka-math.org
G3-M1-SE-B1-1.3.1-01.2016

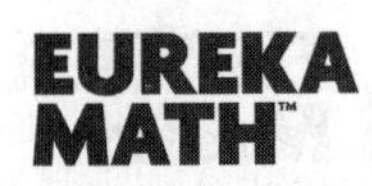

Name ______________________________ Date ______________

1. Label the tape diagrams and complete the equations. Then, draw an array to represent the problems.

a.

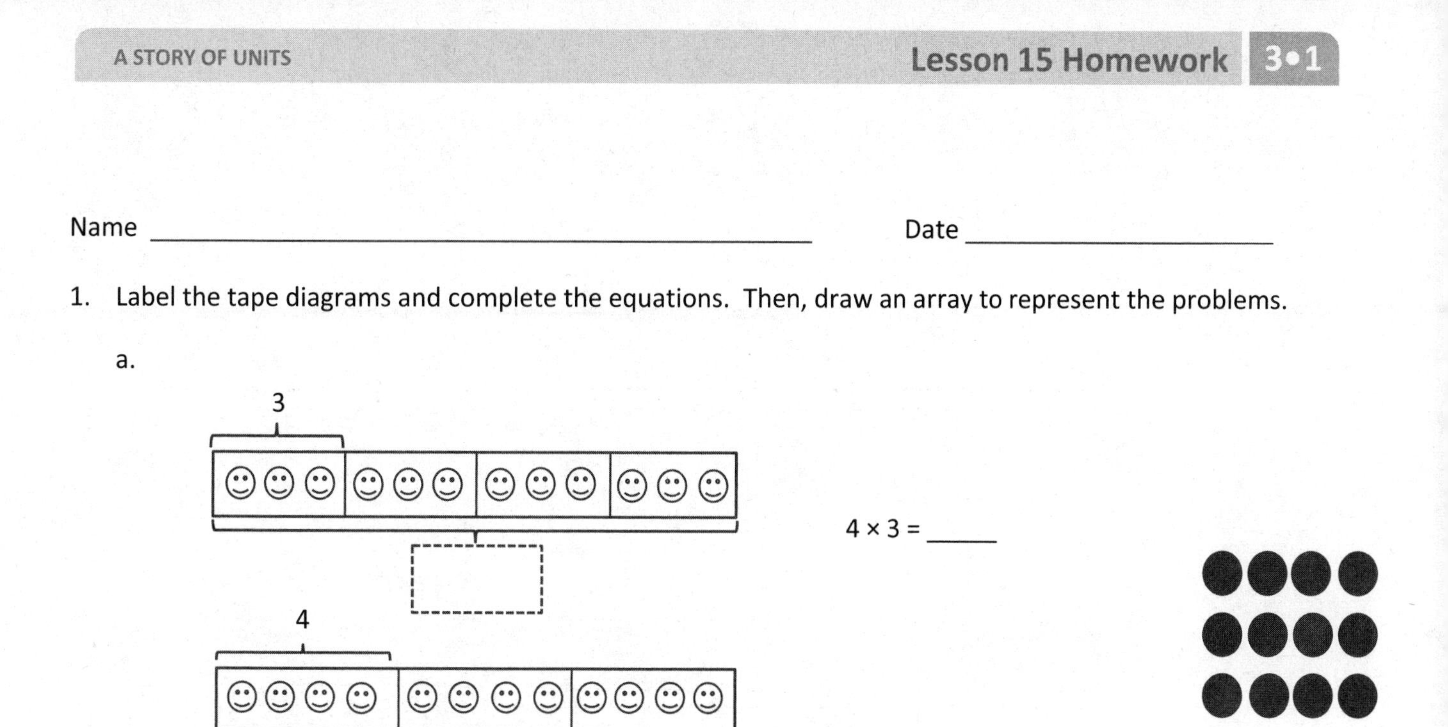

4 × 3 = _____

3 × 4 = _____

b.

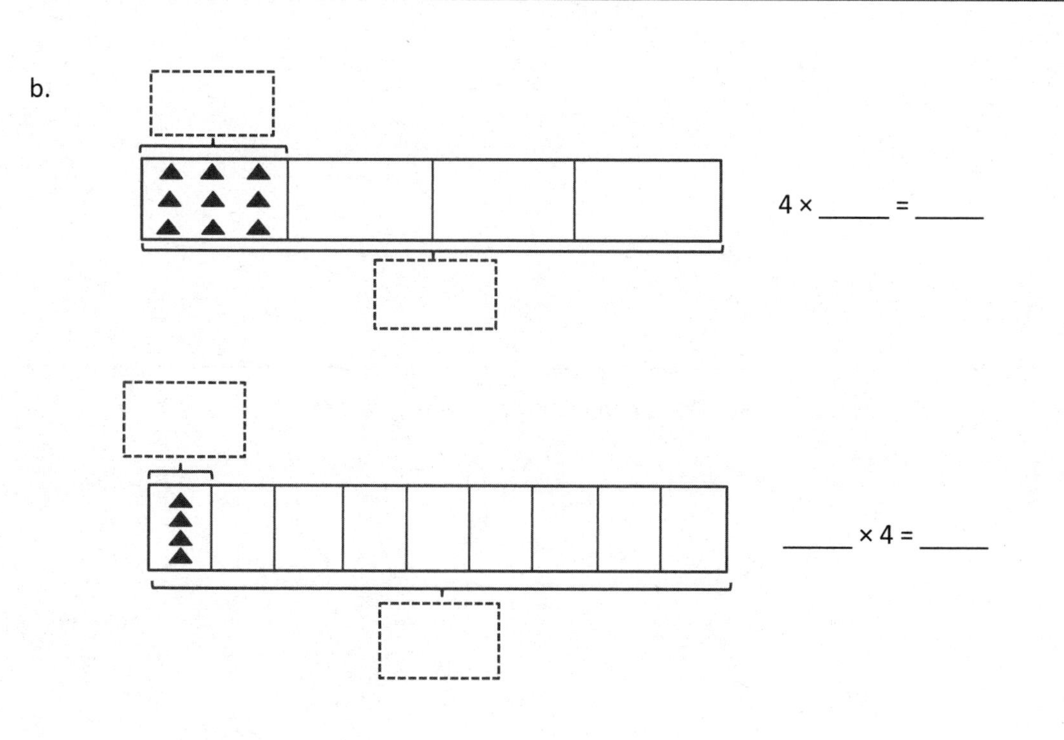

4 × _____ = _____

_____ × 4 = _____

Lesson 15: Relate arrays to tape diagrams to model the commutative property of multiplication.

©2015 Great Minds. eureka-math.org
G3-M1-SE-B1-1.3.1-01.2016

c.

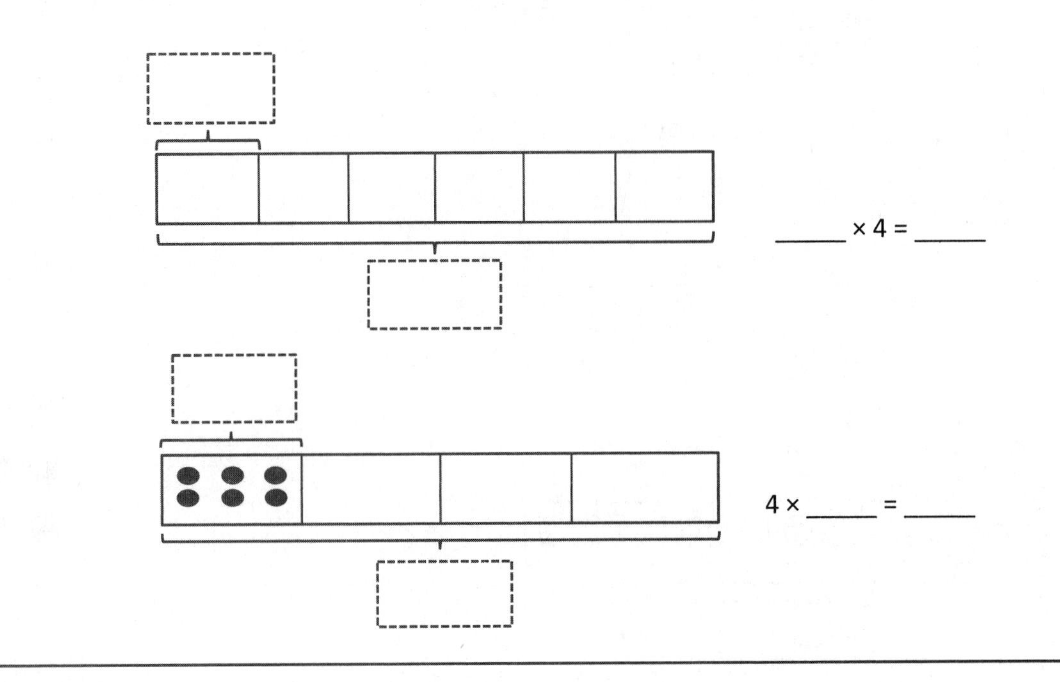

2. Seven clowns hold 4 balloons each at the fair. Draw and label a tape diagram to show the total number of balloons the clowns hold.

3. George swims 7 laps in the pool each day. How many laps does George swim after 4 days?

©2015 Great Minds. eureka-math.org
G3-M1-SE-B1-1.3.1-01.2016

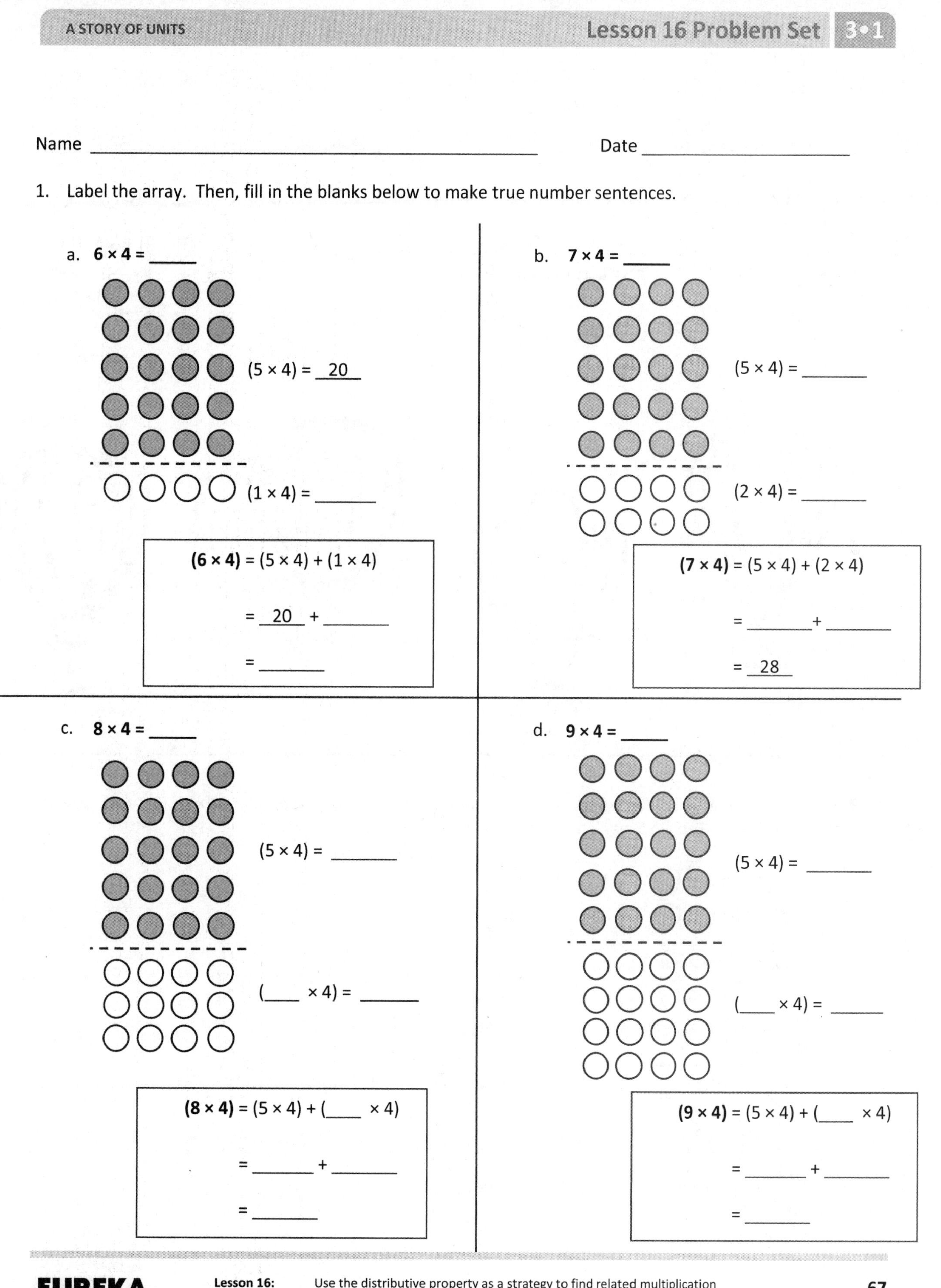

Name ______________________________ Date ______________

1. Label the array. Then, fill in the blanks below to make true number sentences.

a. **6 × 4 =** _____

(5 × 4) = 20

(1 × 4) = ______

(6 × 4) = (5 × 4) + (1 × 4)

= 20 + ______

= ______

b. **7 × 4 =** _____

(5 × 4) = ______

(2 × 4) = ______

(7 × 4) = (5 × 4) + (2 × 4)

= ______ + ______

= 28

c. **8 × 4 =** _____

(5 × 4) = ______

(____ × 4) = ______

(8 × 4) = (5 × 4) + (____ × 4)

= ______ + ______

= ______

d. **9 × 4 =** _____

(5 × 4) = ______

(____ × 4) = ______

(9 × 4) = (5 × 4) + (____ × 4)

= ______ + ______

= ______

©2015 Great Minds. eureka-math.org
G3-M1-SE-B1-1.3.1-01.2016

2. Match the equal expressions.

$(5 \times 4) + (3 \times 4)$

$(5 \times 4) + (1 \times 4)$

$(5 \times 4) + (4 \times 4)$

$(5 \times 4) + (2 \times 4)$

9×4	8×4	6×4	7×4
36	32	24	28

3. Nolan draws the array below to find the answer to the multiplication expression 10×4. He says, "10×4 is just double 5×4." Explain Nolan's strategy.

©2015 Great Minds. eureka-math.org
G3-M1-SE-B1-1.3.1-01.2016

Name ____________________ Date ____________

1. Label the array. Then, fill in the blanks below to make true number sentences.

a. **6 × 4 =** _____

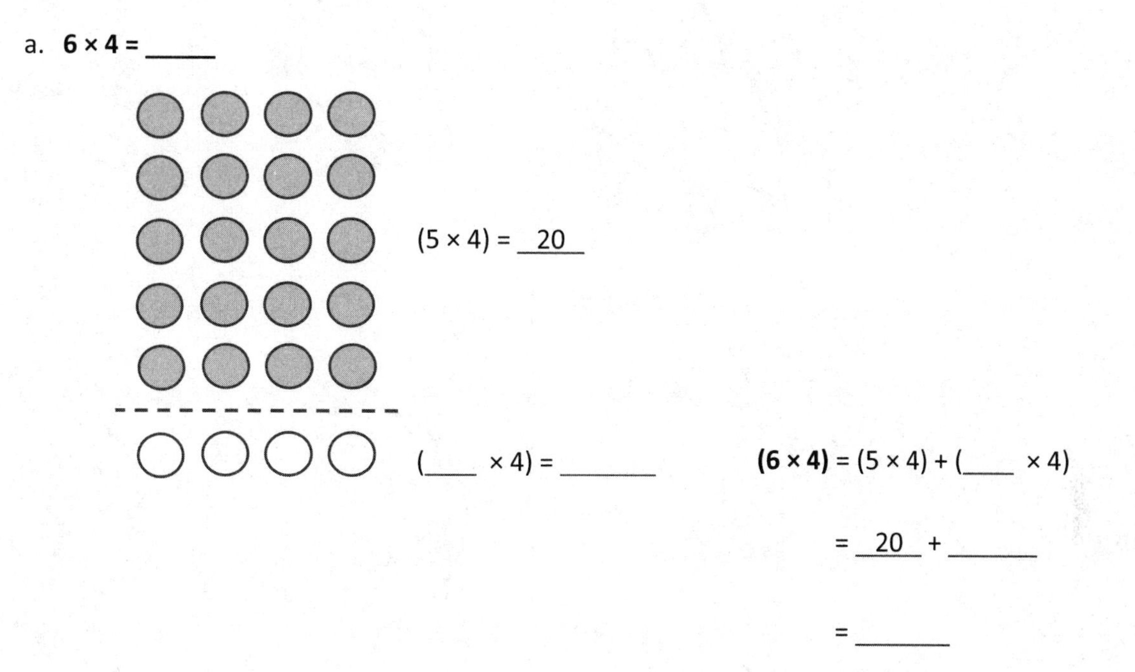

(5 × 4) = 20

(___ × 4) = ______

(6 × 4) = (5 × 4) + (___ × 4)

= 20 + ______

= ______

b. **8 × 4 =** _____

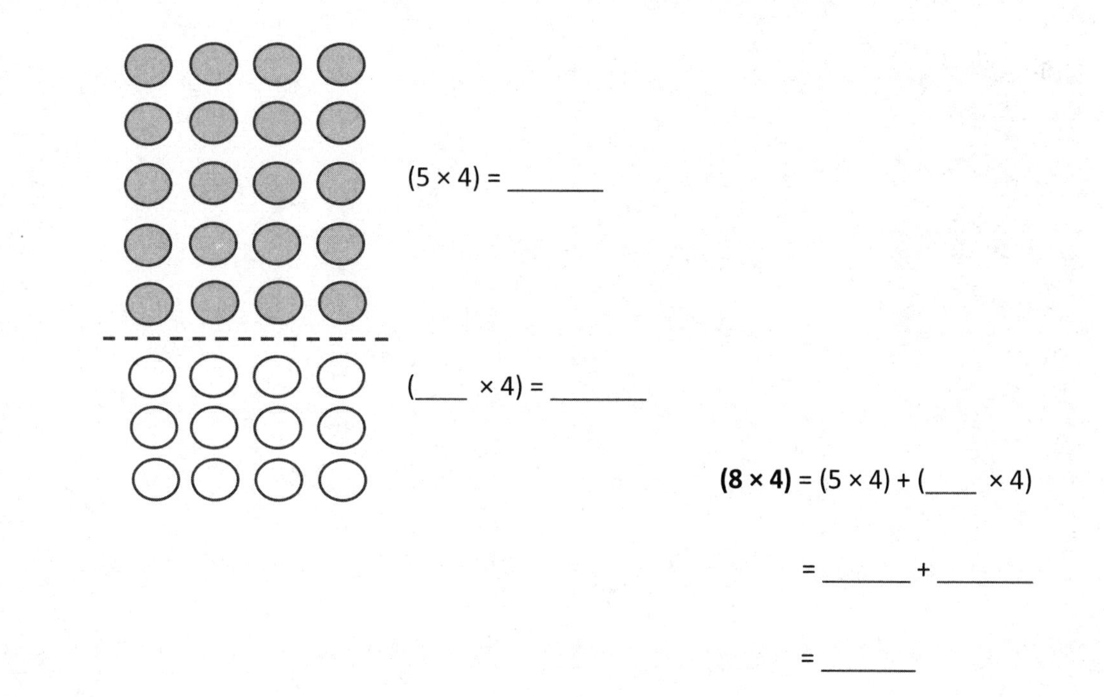

(5 × 4) = ______

(___ × 4) = ______

(8 × 4) = (5 × 4) + (___ × 4)

= ______ + ______

= ______

©2015 Great Minds. eureka-math.org
G3-M1-SE-B1-1.3.1-01.2016

2. Match the multiplication expressions with their answers.

3. The array below shows one strategy for solving 9 × 4. Explain the strategy using your own words.

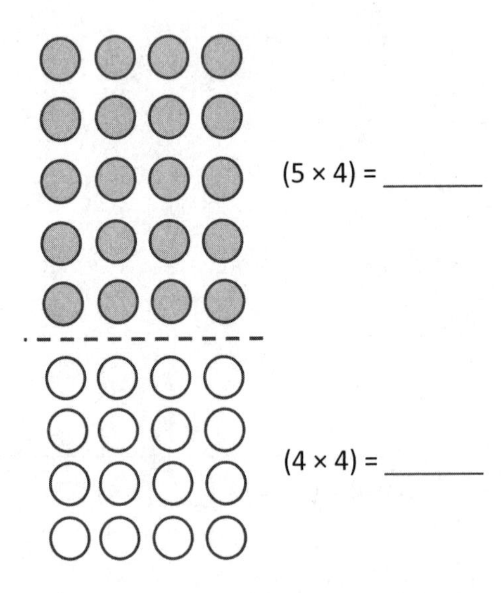

(5 × 4) = ______

(4 × 4) = ______

©2015 Great Minds. eureka-math.org
G3-M1-SE-B1-1.3.1-01.2016

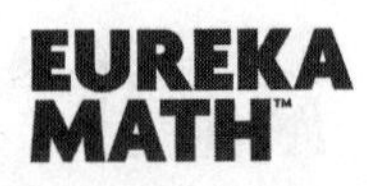

Name ______________________________ Date ______________

1. Use the array to complete the related equations.

$1 \times 4 =$ ______ ______ $\div 4 = 1$

$2 \times 4 =$ ______ ______ $\div 4 = 2$

______ $\times 4 = 12$ $12 \div 4 =$ ______

______ $\times 4 = 16$ $16 \div 4 =$ ______

______ $\times$ ______ $= 20$ $20 \div$ ______ $=$ ______

______ $\times$ ______ $= 24$ $24 \div$ ______ $=$ ______

______ $\times 4 =$ ______ ______ $\div 4 =$ ______

______ $\times 4 =$ ______ ______ $\div 4 =$ ______

______ $\times$ ______ $=$ ______ ______ $\div$ ______ $=$ ______

______ $\times$ ______ $=$ ______ ______ $\div$ ______ $=$ ______

©2015 Great Minds. eureka-math.org
G3-M1-SE-B1-1.3.1-01.2016

2. The teacher puts 32 students into groups of 4. How many groups does she make? Draw and label a tape diagram to solve.

---

3. The store clerk arranges 24 toothbrushes into 4 equal rows. How many toothbrushes are in each row?

---

4. An art teacher has 40 paintbrushes. She divides them equally among her 4 students. She finds 8 more brushes and divides these equally among the students, as well. How many brushes does each student receive?

©2015 Great Minds. eureka-math.org
G3-M1-SE-B1-1.3.1-01.2016

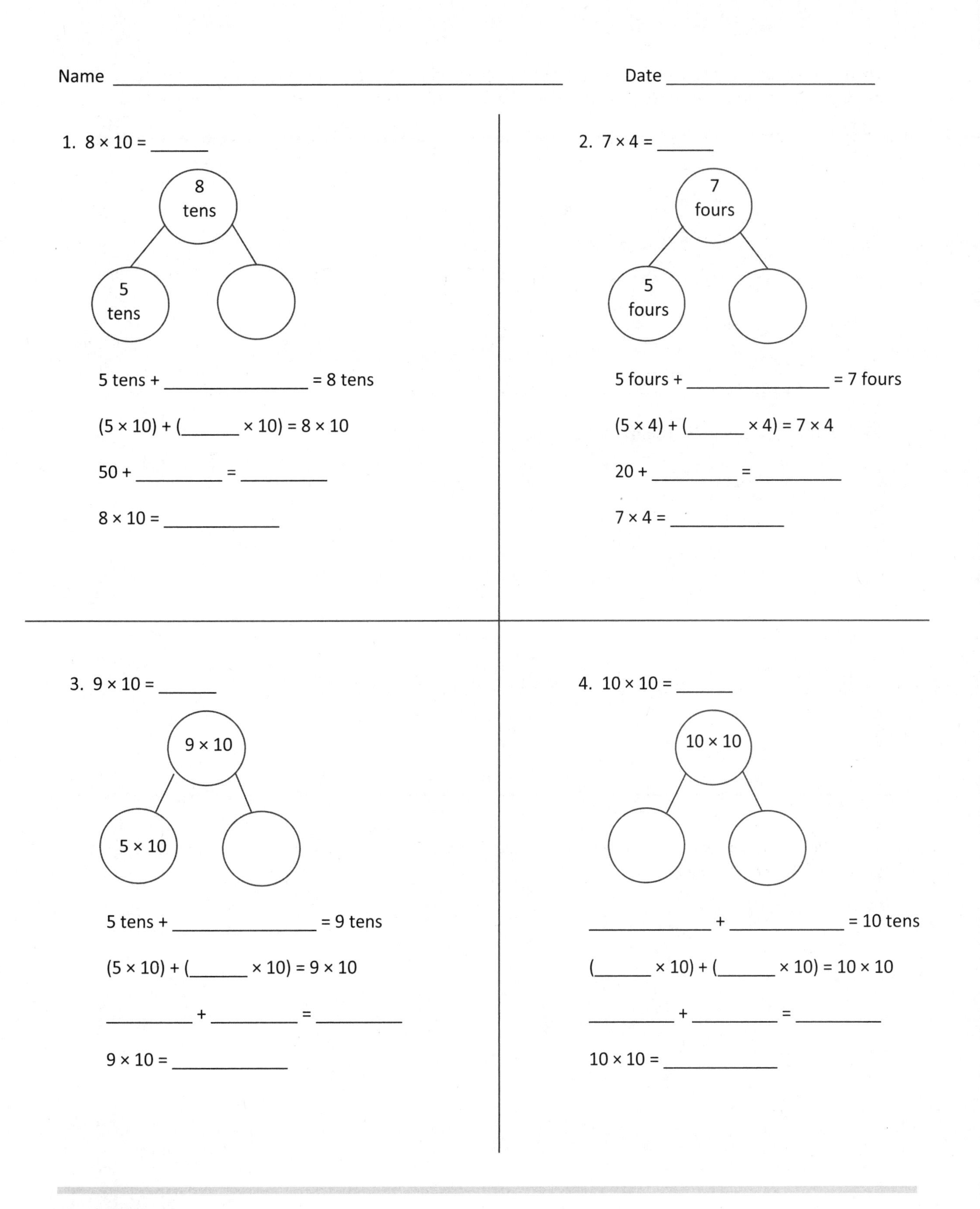

Name ______________________________ Date ________________

1. 8 × 10 = ______

5 tens + _______________ = 8 tens

(5 × 10) + (______ × 10) = 8 × 10

50 + _________ = _________

8 × 10 = ____________

2. 7 × 4 = ______

5 fours + _______________ = 7 fours

(5 × 4) + (______ × 4) = 7 × 4

20 + _________ = _________

7 × 4 = ____________

3. 9 × 10 = ______

5 tens + _______________ = 9 tens

(5 × 10) + (______ × 10) = 9 × 10

_________ + _________ = _________

9 × 10 = ____________

4. 10 × 10 = ______

_____________ + ____________ = 10 tens

(______ × 10) + (______ × 10) = 10 × 10

_________ + _________ = _________

10 × 10 = ____________

©2015 Great Minds. eureka-math.org
G3-M1-SE-B1-1.3.1-01.2016

5. There are 7 teams in the soccer tournament. Ten children play on each team. How many children are playing in the tournament? Use the break apart and distribute strategy, and draw a number bond to solve.

There are __________ children playing in the tournament.

6. What is the total number of sides on 8 triangles?

7. There are 12 rows of bottled drinks in the vending machine. Each row has 10 bottles. How many bottles are in the vending machine?

©2015 Great Minds. eureka-math.org
G3-M1-SE-B1-1.3.1-01.2016

Name ______________________________ Date ________________

1. Match.

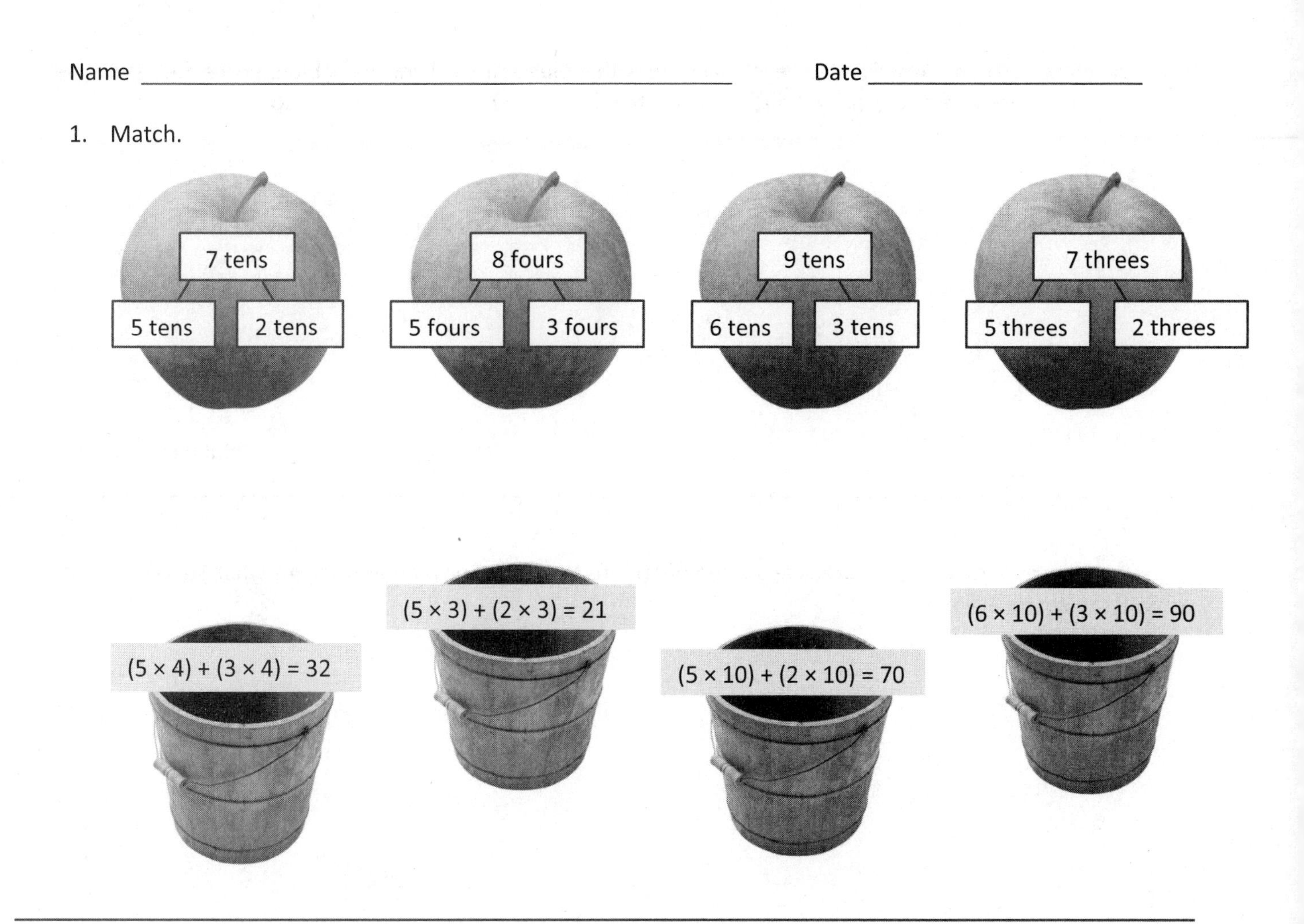

2. 9 × 4 = _______

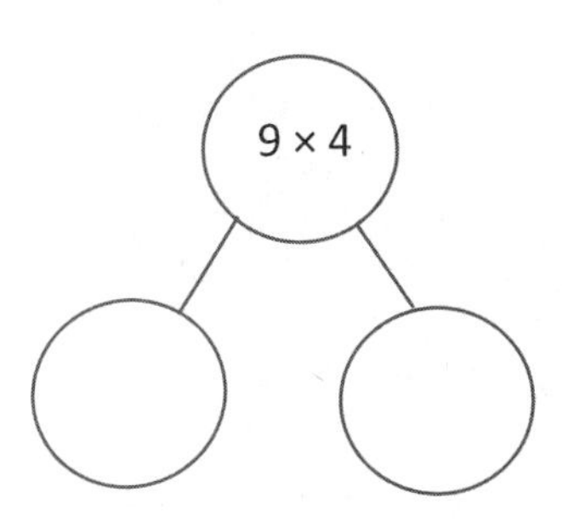

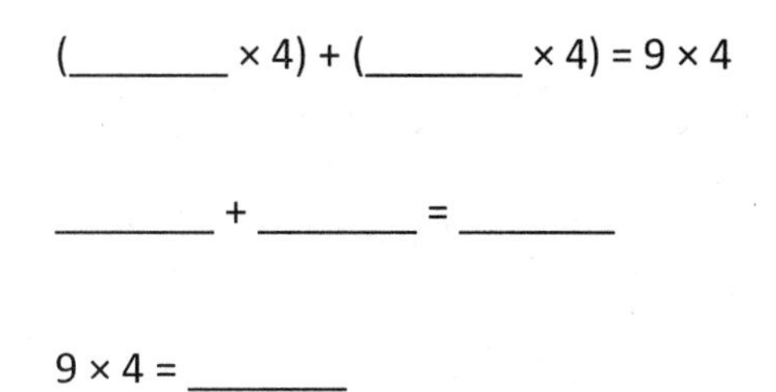

©2015 Great Minds. eureka-math.org
G3-M1-SE-B1-1.3.1-01.2016

3. Lydia makes 10 pancakes. She tops each pancake with 4 blueberries. How many blueberries does Lydia use in all? Use the break apart and distribute strategy, and draw a number bond to solve.

Lydia uses _______ blueberries in all.

---

4. Steven solves 7 × 3 using the break apart and distribute strategy. Show an example of what Steven's work might look like below.

---

5. There are 7 days in 1 week. How many days are there in 10 weeks?

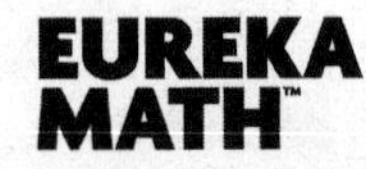

©2015 Great Minds. eureka-math.org
G3-M1-SE-B1-1.3.1-01.2016

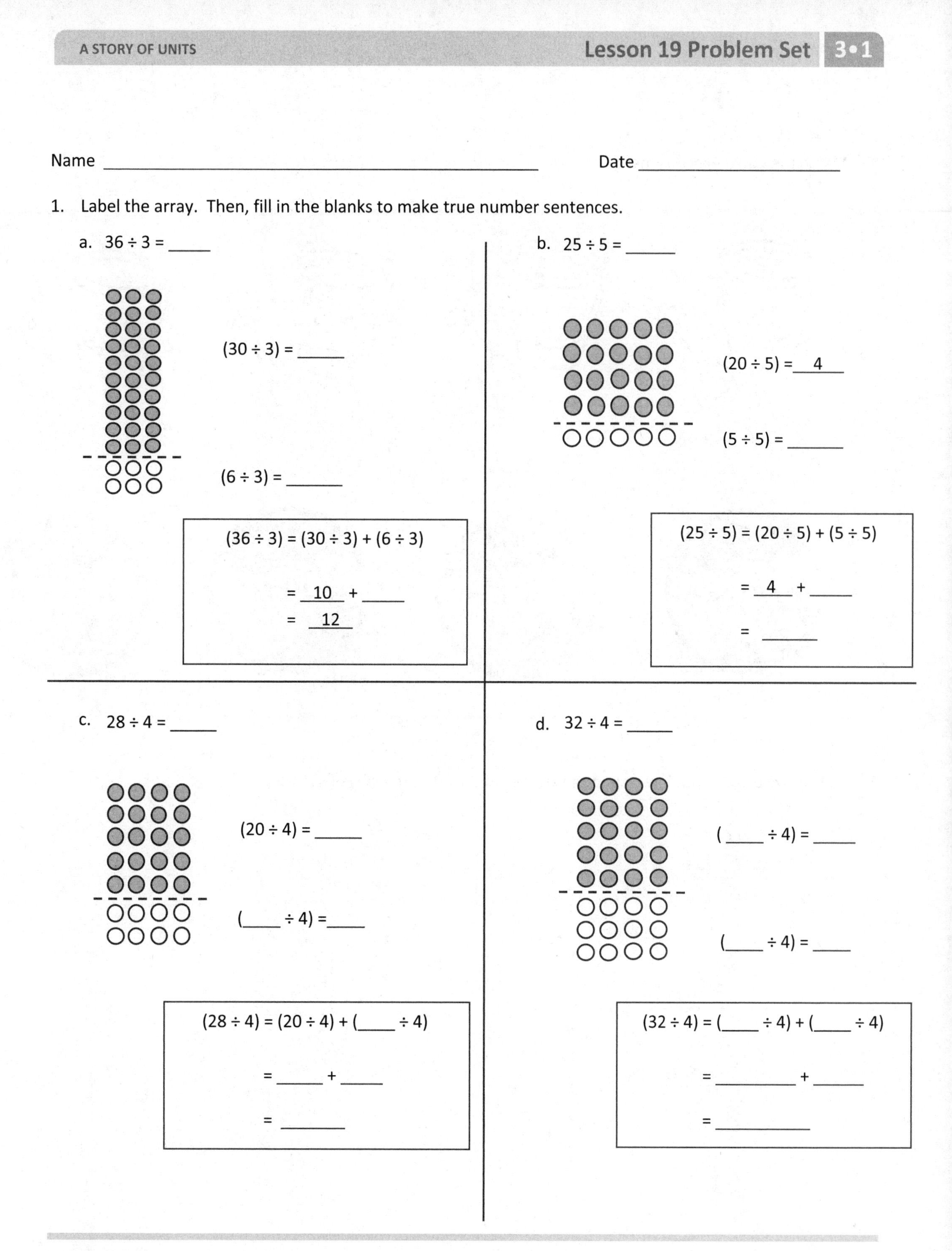

Name ____________________ Date ____________

1. Label the array. Then, fill in the blanks to make true number sentences.

a. $36 \div 3 =$ ____

$(30 \div 3) =$ ____

$(6 \div 3) =$ ____

$(36 \div 3) = (30 \div 3) + (6 \div 3)$

$=$ 10 + ____

$=$ 12

b. $25 \div 5 =$ ____

$(20 \div 5) =$ 4

$(5 \div 5) =$ ____

$(25 \div 5) = (20 \div 5) + (5 \div 5)$

$=$ 4 + ____

$=$ ____

c. $28 \div 4 =$ ____

$(20 \div 4) =$ ____

(____ $\div 4) =$ ____

$(28 \div 4) = (20 \div 4) +$ (____ $\div 4)$

$=$ ____ + ____

$=$ ____

d. $32 \div 4 =$ ____

(____ $\div 4) =$ ____

(____ $\div 4) =$ ____

$(32 \div 4) =$ (____ $\div 4) +$ (____ $\div 4)$

$=$ ____ + ____

$=$ ____

©2015 Great Minds. eureka-math.org
G3-M1-SE-B1-1.3.1-01.2016

2. Match the equal expressions.

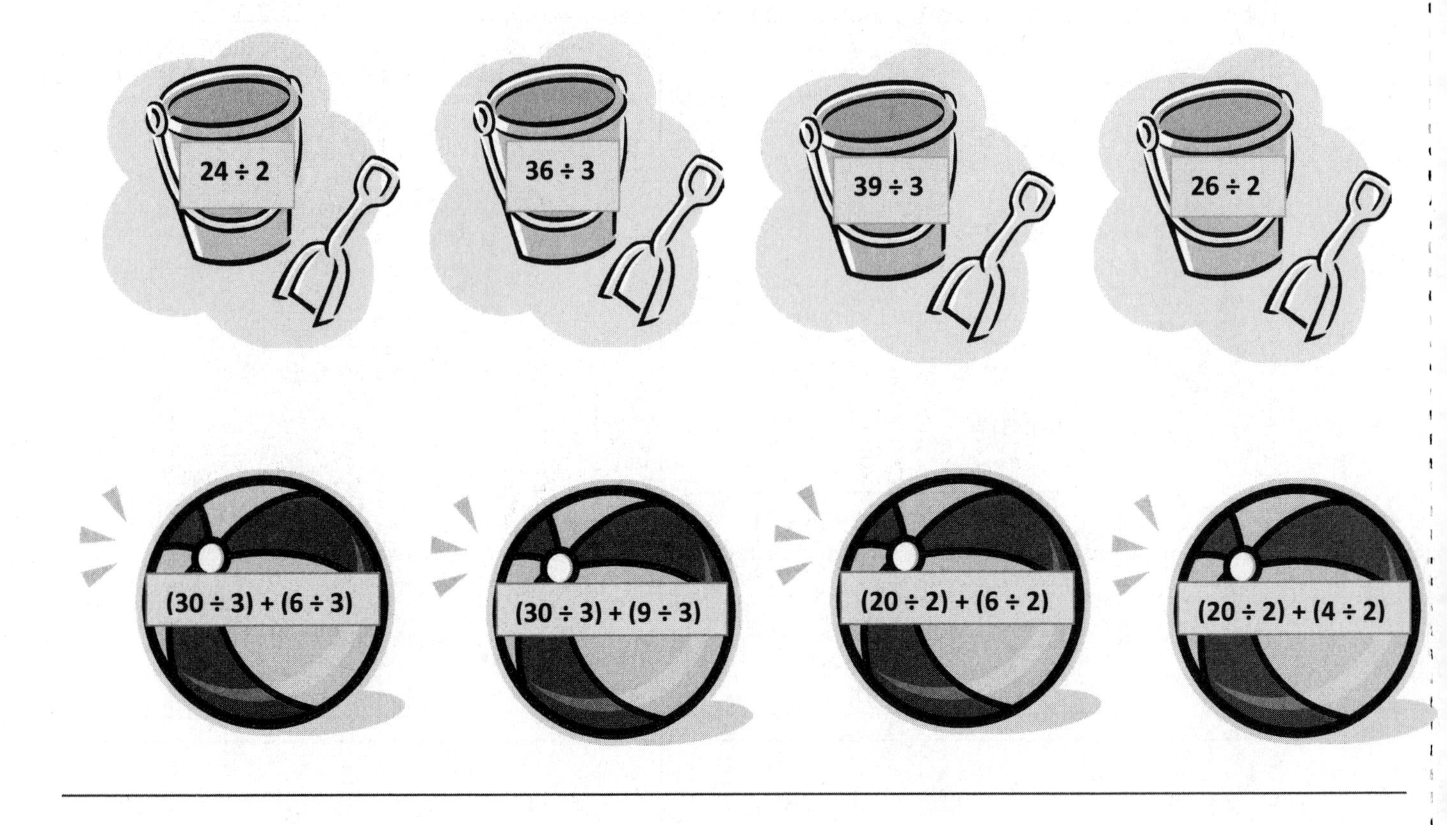

3. Nell draws the array below to find the answer to 24 ÷ 2. Explain Nell's strategy.

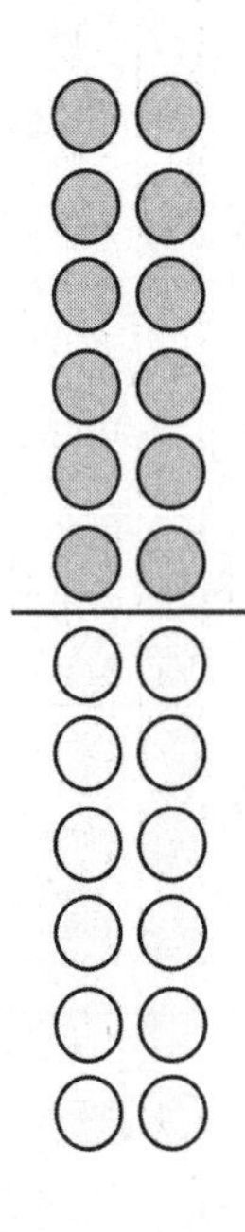

©2015 Great Minds. eureka-math.org
G3-M1-SE-B1-1.3.1-01.2016

Name _____ Date _____

1. Label the array. Then, fill in the blanks to make true number sentences.

a. $18 \div 3 =$ _____

$(9 \div 3) = 3$

$(9 \div 3) =$ _____

$(18 \div 3) = (9 \div 3) + (9 \div 3)$

$= \underline{\ 3\ } + \underline{\hspace{1cm}}$

$= \underline{\ 6\ }$

b. $21 \div 3 =$ _____

$(15 \div 3) = 5$

$(6 \div 3) =$ _____

$(21 \div 3) = (15 \div 3) + (6 \div 3)$

$= \underline{\ 5\ } + \underline{\hspace{1cm}}$

$= \underline{\hspace{1cm}}$

c. $24 \div 4 =$ _____

$(20 \div 4) =$ _____

$(4 \div 4) =$ _____

$(24 \div 4) = (20 \div 4) + (\underline{\hspace{0.5cm}} \div 4)$

$= \underline{\hspace{1cm}} + \underline{\hspace{1cm}}$

$= \underline{\hspace{1cm}}$

d. $36 \div 4 =$ _____

$(20 \div 4) =$ _____

$(16 \div 4) =$ _____

$(36 \div 4) = (\underline{\hspace{0.5cm}} \div 4) + (\underline{\hspace{0.5cm}} \div 4)$

$= \underline{\hspace{1cm}} + \underline{\hspace{1cm}}$

$= \underline{\hspace{1cm}}$

©2015 Great Minds. eureka-math.org
G3-M1-SE-B1-1.3.1-01.2016

2. Match equal expressions.

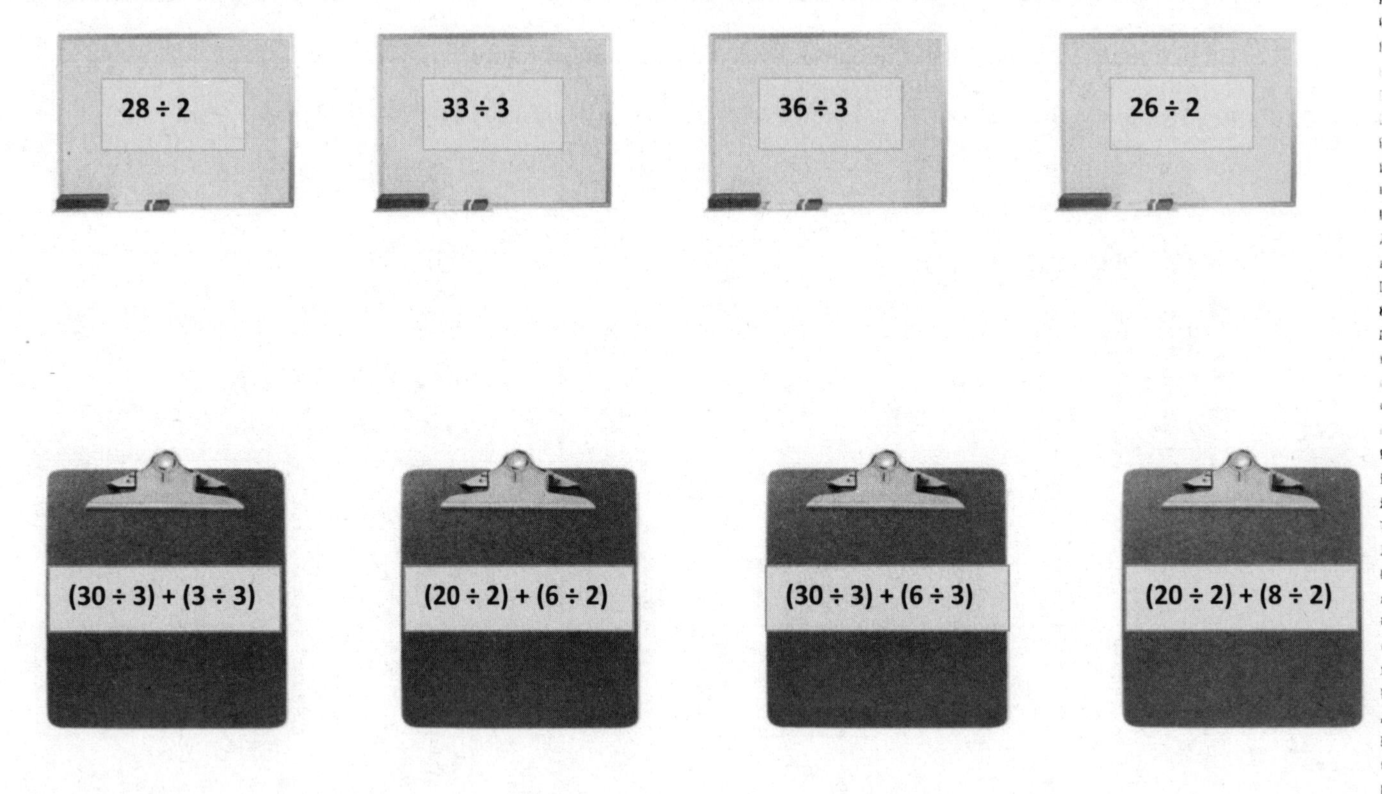

3. Alex draws the array below to find the answer to 35 ÷ 5. Explain Alex's strategy.

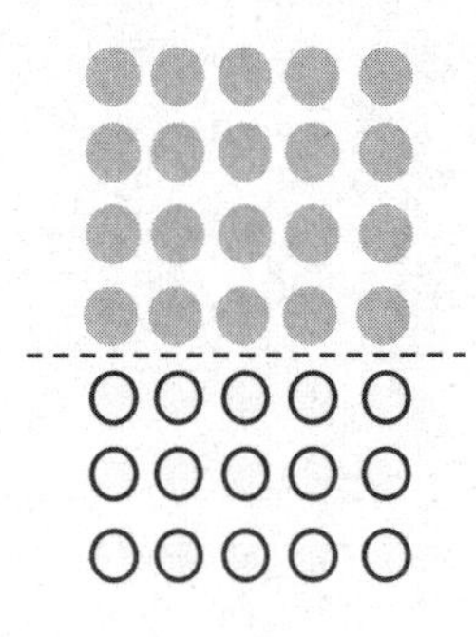

©2015 Great Minds. eureka-math.org
G3-M1-SE-B1-1.3.1-01.2016

Name ______________________________ Date ________________

1. Ted buys 3 books and a magazine at the book store. Each book costs $8. A magazine costs $4.

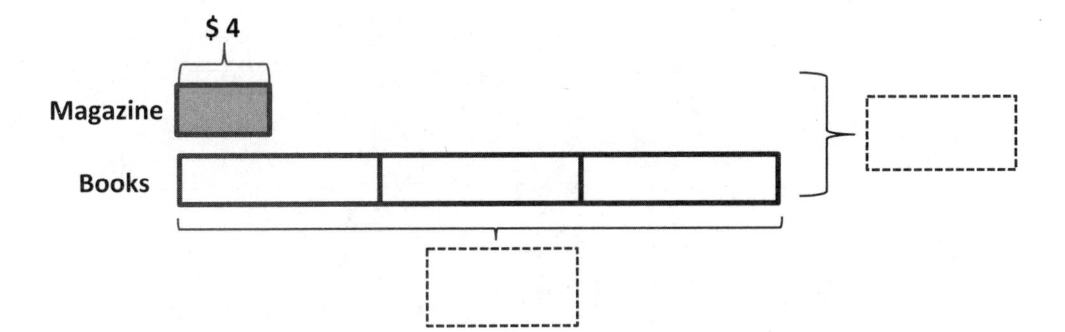

a. What is the total cost of the books?

b. How much does Ted spend altogether?

---

2. Seven children share 28 silly bands equally.

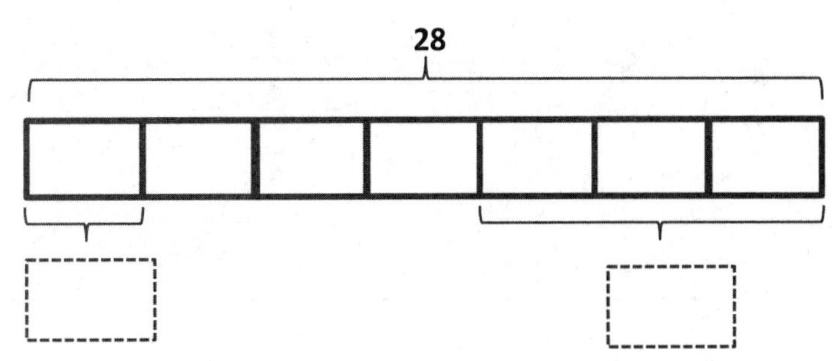

a. How many silly bands does each child get?

b. How many silly bands do 3 children get?

©2015 Great Minds. eureka-math.org
G3-M1-SE-B1-1.3.1-01.2016

3. Eighteen cups are equally packed into 6 boxes. Two boxes of cups break. How many cups are unbroken?

---

4. There are 25 blue balloons and 15 red balloons at a party. Five children are given an equal number of each color balloon. How many blue and red balloons does each child get?

---

5. Twenty-seven pears are packed in bags of 3. Five bags of pears are sold. How many bags of pears are left?

©2015 Great Minds. eureka-math.org
G3-M1-SE-B1-1.3.1-01.2016

Name ________________________________ Date ________________

1. Jerry buys a pack of pencils that costs $3. David buys 4 sets of markers. Each set of markers also costs $3.

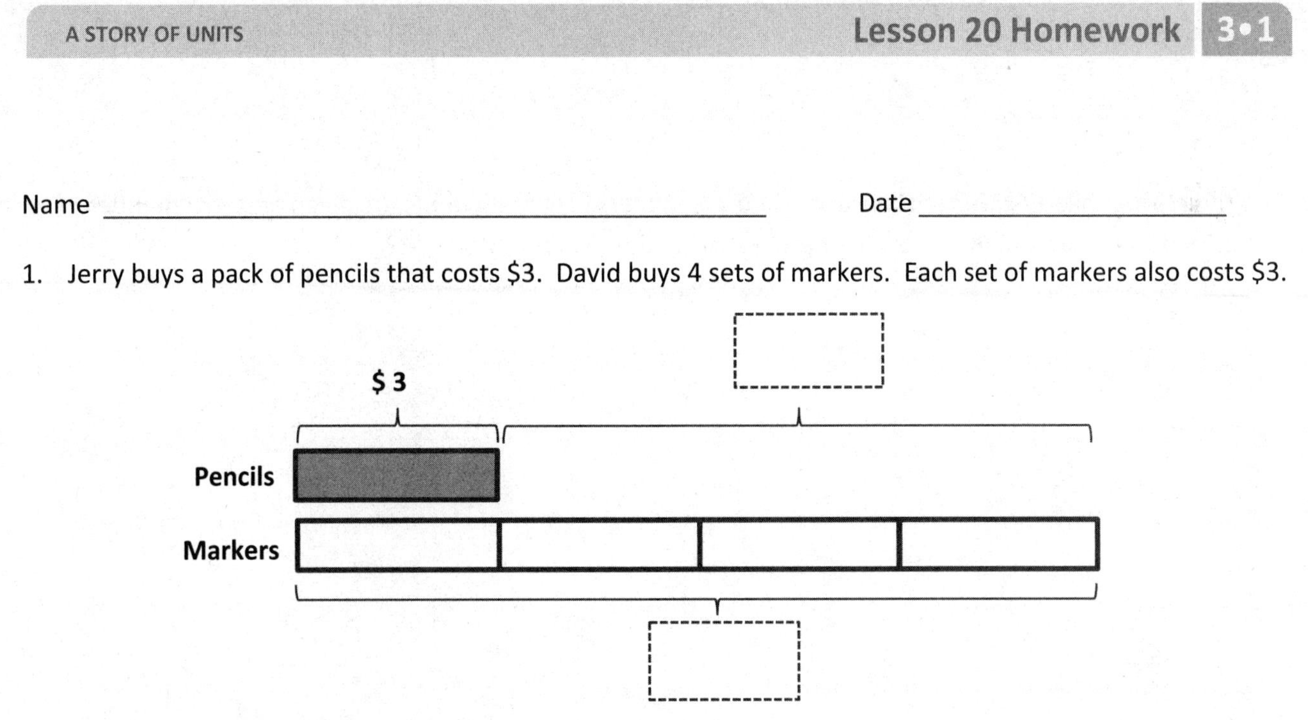

   a. What is the total cost of the markers?

   b. How much more does David spend on 4 sets of markers than Jerry spends on a pack of pencils?

---

2. Thirty students are eating lunch at 5 tables. Each table has the same number of students.

   a. How many students are sitting at each table?

   30 children

   b. How many students are sitting at 4 tables?

©2015 Great Minds. eureka-math.org
G3-M1-SE-B1-1.3.1-01.2016

3. The teacher has 12 green stickers and 15 purple stickers. Three students are given an equal number of each color sticker. How many green and purple stickers does each student get?

---

4. Three friends go apple picking. They pick 13 apples on Saturday and 14 apples on Sunday. They share the apples equally. How many apples does each person get?

---

5. The store has 28 notebooks in packs of 4. Three packs of notebooks are sold. How many packs of notebooks are left?

©2015 Great Minds. eureka-math.org
G3-M1-SE-B1-1.3.1-01.2016

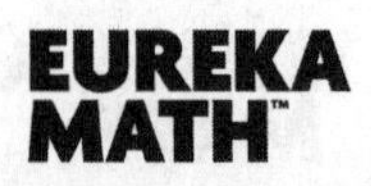

Name ________________________________ Date ________________

1. Jason earns $6 per week for doing all his chores. On the fifth week, he forgets to take out the trash, so he only earns $4. Write and solve an equation to show how much Jason earns in 5 weeks.

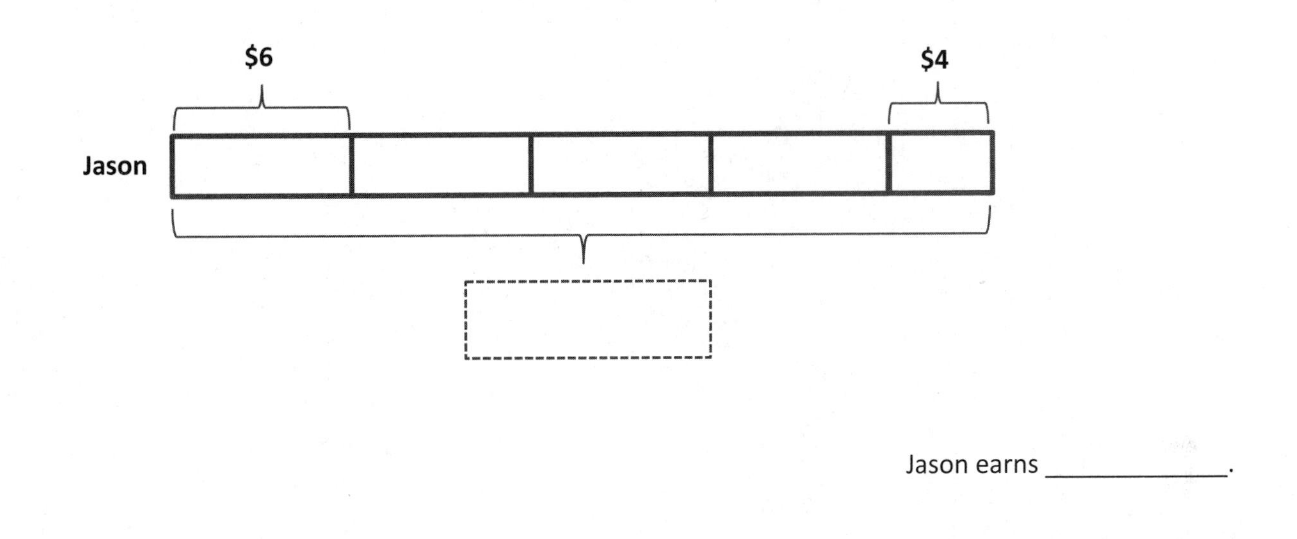

Jason earns __________.

2. Miss Lianto orders 4 packs of 7 markers. After passing out 1 marker to each student in her class, she has 6 left. Label the tape diagram to find how many students are in Miss Lianto's class.

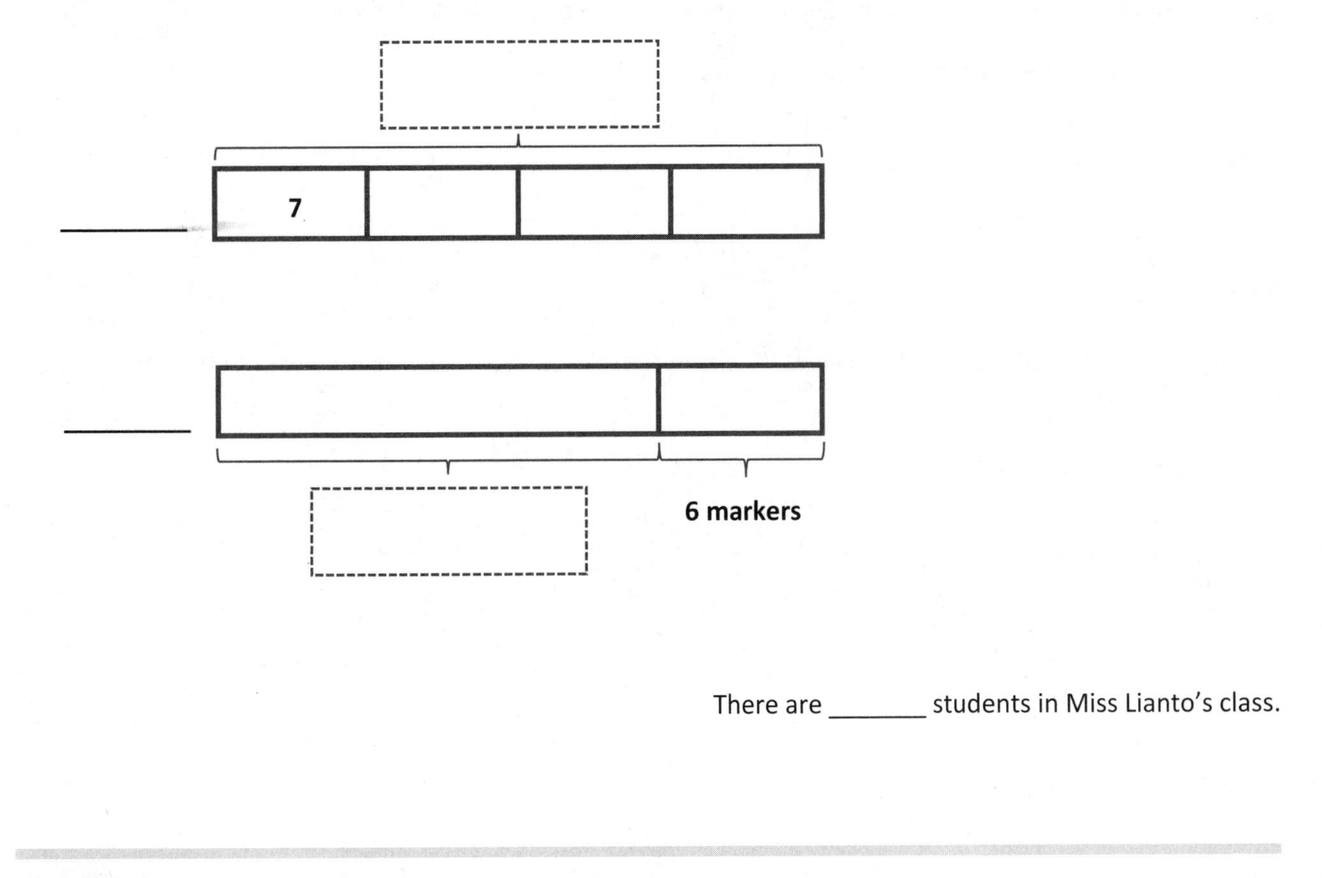

There are ______ students in Miss Lianto's class.

©2015 Great Minds. eureka-math.org
G3-M1-SE-B1-1.3.1-01.2016

3. Orlando buys a box of 18 fruit snacks. Each box comes with an equal number of strawberry-, cherry-, and grape-flavored snacks. He eats all of the grape-flavored snacks. Draw and label a tape diagram to find how many fruit snacks he has left.

---

4. Eudora buys 21 meters of ribbon. She cuts the ribbon so that each piece measures 3 meters in length.

   a. How many pieces of ribbon does she have?

   b. If Eudora needs a total of 12 pieces of the shorter ribbon, how many more pieces of the shorter ribbon does she need?

©2015 Great Minds. eureka-math.org
G3-M1-SE-B1-1.3.1-01.2016

Name ______________________________ Date ________________

1. Tina eats 8 crackers for a snack each day at school. On Friday, she drops 3 and only eats 5. Write and solve an equation to show the total number of crackers Tina eats during the week.

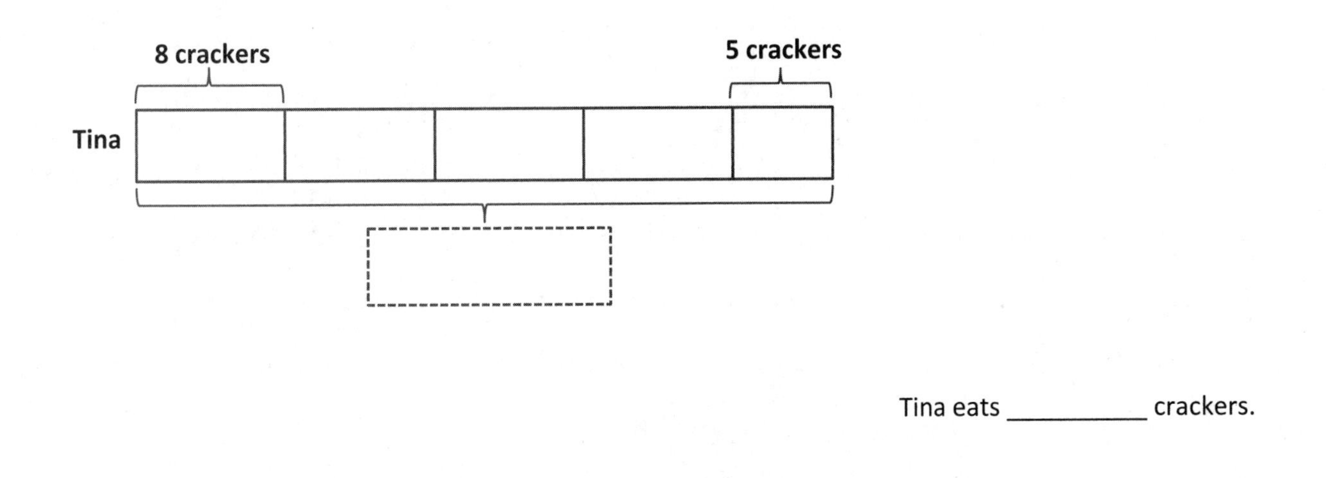

Tina eats __________ crackers.

---

2. Ballio has a reading goal. He checks 3 boxes of 9 books out from the library. After finishing them, he realizes that he beat his goal by 4 books! Label the tape diagrams to find Ballio's reading goal.

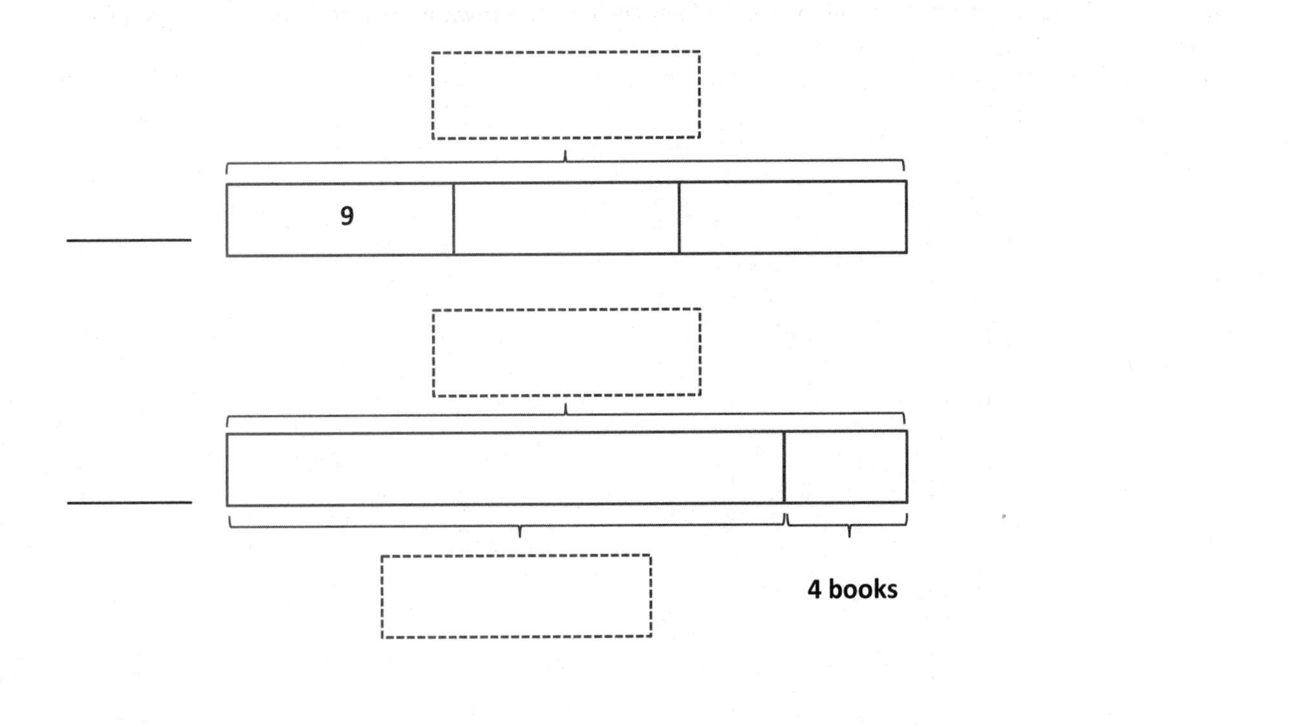

Ballio's goal is to read _______ books.

©2015 Great Minds. eureka-math.org
G3-M1-SE-B1-1.3.1-01.2016

3. Mr. Nguyen plants 24 trees around the neighborhood pond. He plants equal numbers of maple, pine, spruce, and birch trees. He waters the spruce and birch trees before it gets dark. How many trees does Mr. Nguyen still need to water? Draw and label a tape diagram.

---

4. Anna buys 24 seeds and plants 3 in each pot. She has 5 pots. How many more pots does Anna need to plant all of her seeds?

©2015 Great Minds. eureka-math.org
G3-M1-SE-B1-1.3.1-01.2016

Student Edition

Eureka Math

Grade 3
Module 2

Special thanks go to the Gordon A. Cain Center and to the Department of Mathematics at Louisiana State University for their support in the development of *Eureka Math.*

Published by the non-profit Great Minds

Copyright © 2015 Great Minds. No part of this work may be reproduced, sold, or commercialized, in whole or in part, without written permission from Great Minds. Non-commercial use is licensed pursuant to a Creative Commons Attribution-NonCommercial-ShareAlike 4.0 license; for more information, go to http://greatminds.net/maps/math/copyright. "Great Minds" and "Eureka Math" are registered trademarks of Great Minds.

Printed in the U.S.A.
This book may be purchased from the publisher at eureka-math.org
10 9 8 7 6

Name ______________________________ Date ______________

1. Use a stopwatch. How long does it take you to snap your fingers 10 times?

It takes ______________________ to snap 10 times.

2. Use a stopwatch. How long does it take to write every whole number from 0 to 25?

It takes ______________________ to write every whole number from 0 to 25.

3. Use a stopwatch. How long does it take you to name 10 animals? Record them below.

It takes ______________________ to name 10 animals.

4. Use a stopwatch. How long does it take you to write $7 \times 8 = 56$ fifteen times? Record the time below.

It takes ______________________ to write $7 \times 8 = 56$ fifteen times.

©2015 Great Minds. eureka-math.org
G3-M2-SE-B1-1.3.1-01.2016

5. Work with your group. Use a stopwatch to measure the time for each of the following activities.

Activity	Time
Write your full name.	____________ seconds
Do 20 jumping jacks.	
Whisper count by twos from 0 to 30.	
Draw 8 squares.	
Skip-count out loud by fours from 24 to 0.	
Say the names of your teachers from Kindergarten to Grade 3.	

6. 100 meter relay: Use a stopwatch to measure and record your team's times.

Name	Time
	Total time:

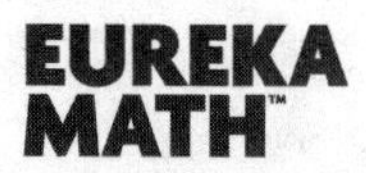

©2015 Great Minds. eureka-math.org
G3-M2-SE-B1-1.3.1-01.2016

Name ______________________________ Date ______________

1. The table to the right shows how much time it takes each of the 5 students to run 100 meters.

Samantha	19 seconds
Melanie	22 seconds
Chester	26 seconds
Dominique	18 seconds
Louie	24 seconds

   a. Who is the fastest runner?

   b. Who is the slowest runner?

   c. How many seconds faster did Samantha run than Louie?

2. List activities at home that take about the following amounts of time to complete. If you do not have a stopwatch, you can use the strategy of counting by *1 Mississippi, 2 Mississippi, 3 Mississippi, ….*

Time	Activities at home
30 seconds	Example: Tying shoelaces
45 seconds	
60 seconds	

©2015 Great Minds. eureka-math.org
G3-M2-SE-B1-1.3.1-01.2016

3. Match the analog clock with the correct digital clock.

07:05

11:00

10:15

02:50

©2015 Great Minds. eureka-math.org
G3-M2-SE-B1-1.3.1-01.2016

Name ______________________________ Date ______________

1. Follow the directions to label the number line below.

a. Ingrid gets ready for school between 7:00 a.m. and 8:00 a.m. Label the first and last tick marks as 7:00 a.m. and 8:00 a.m.

b. Each interval represents 5 minutes. Count by fives starting at 0, or 7:00 a.m. Label each 5-minute interval below the number line up to 8:00 a.m.

c. Ingrid starts getting dressed at 7:10 a.m. Plot a point on the number line to represent this time. Above the point, write *D*.

d. Ingrid starts eating breakfast at 7:35 a.m. Plot a point on the number line to represent this time. Above the point, write *E*.

e. Ingrid starts brushing her teeth at 7:40 a.m. Plot a point on the number line to represent this time. Above the point, write *T*.

f. Ingrid starts packing her lunch at 7:45 a.m. Plot a point on the number line to represent this time. Above the point, write *L*.

g. Ingrid starts waiting for the bus at 7:55 a.m. Plot a point on the number line to represent this time. Above the point, write *W*.

©2015 Great Minds. eureka-math.org
G3-M2-SE-B1-1.3.1-01.2016

2. Label every 5 minutes below the number line shown. Draw a line from each clock to the point on the number line which shows its time. Not all of the clocks have matching points.

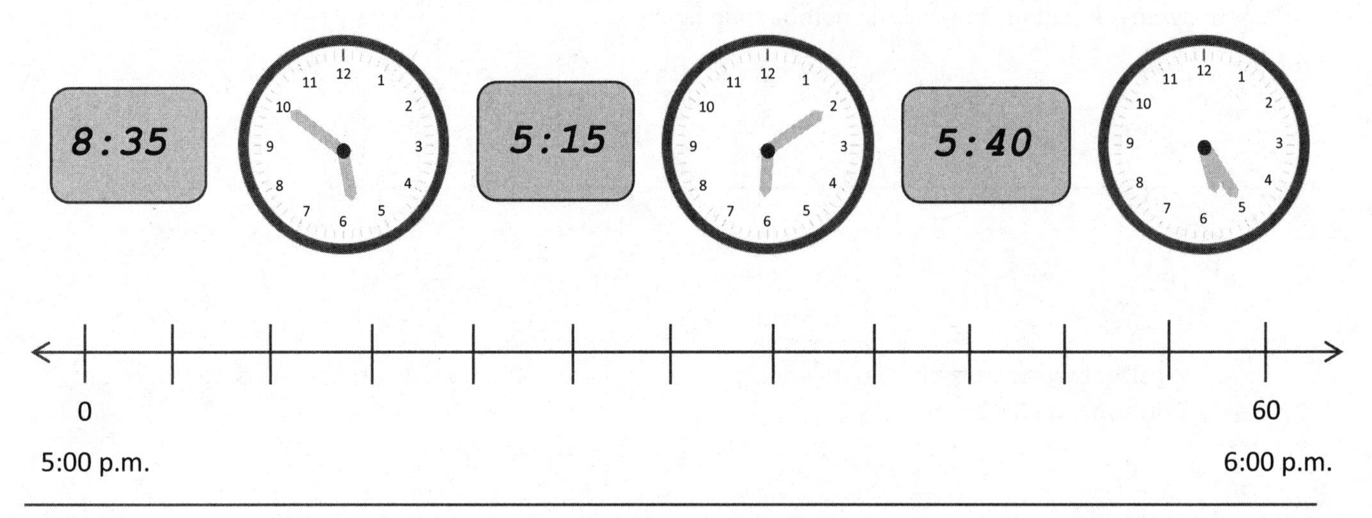

3. Noah uses a number line to locate 5:45 p.m. Each interval is 5 minutes. The number line shows the hour from 5 p.m. to 6 p.m. Label the number line below to show his work.

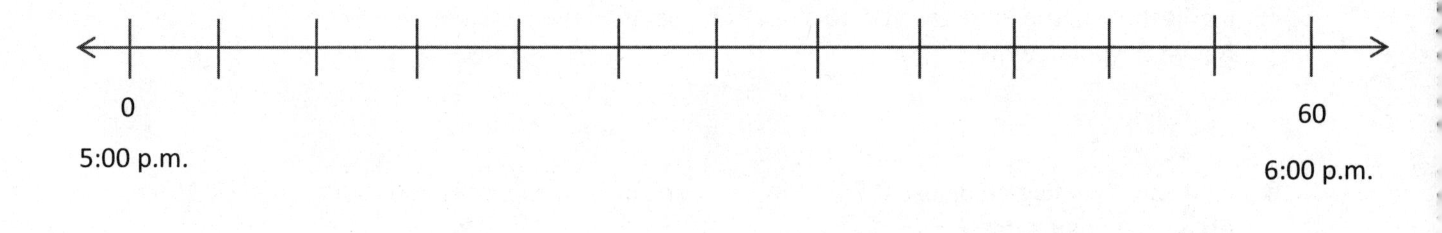

4. Tanner tells his little brother that 11:25 p.m. comes after 11:20 a.m. Do you agree with Tanner? Why or why not?

©2015 Great Minds. eureka-math.org
G3-M2-SE-B1-1.3.1-01.2016

Name ______________________________ Date ________________

Follow the directions to label the number line below.

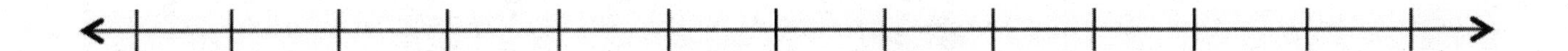

a. The basketball team practices between 4:00 p.m. and 5:00 p.m. Label the first and last tick marks as 4:00 p.m. and 5:00 p.m.

b. Each interval represents 5 minutes. Count by fives starting at 0, or 4:00 p.m. Label each 5-minute interval below the number line up to 5:00 p.m.

c. The team warms up at 4:05 p.m. Plot a point on the number line to represent this time. Above the point, write *W*.

d. The team shoots free throws at 4:15 p.m. Plot a point on the number line to represent this time. Above the point, write *F*.

e. The team plays a practice game at 4:25 p.m. Plot a point on the number line to represent this time. Above the point, write *G*.

f. The team has a water break at 4:50 p.m. Plot a point on the number line to represent this time. Above the point, write *B*.

g. The team reviews their plays at 4:55 p.m. Plot a point on the number line to represent this time. Above the point, write *P*.

©2015 Great Minds. eureka-math.org
G3-M2-SE-B1-1.3.1-01.2016

tape diagram

©2015 Great Minds. eureka-math.org
G3-M2-SE-B1-1.3.1-01.2016

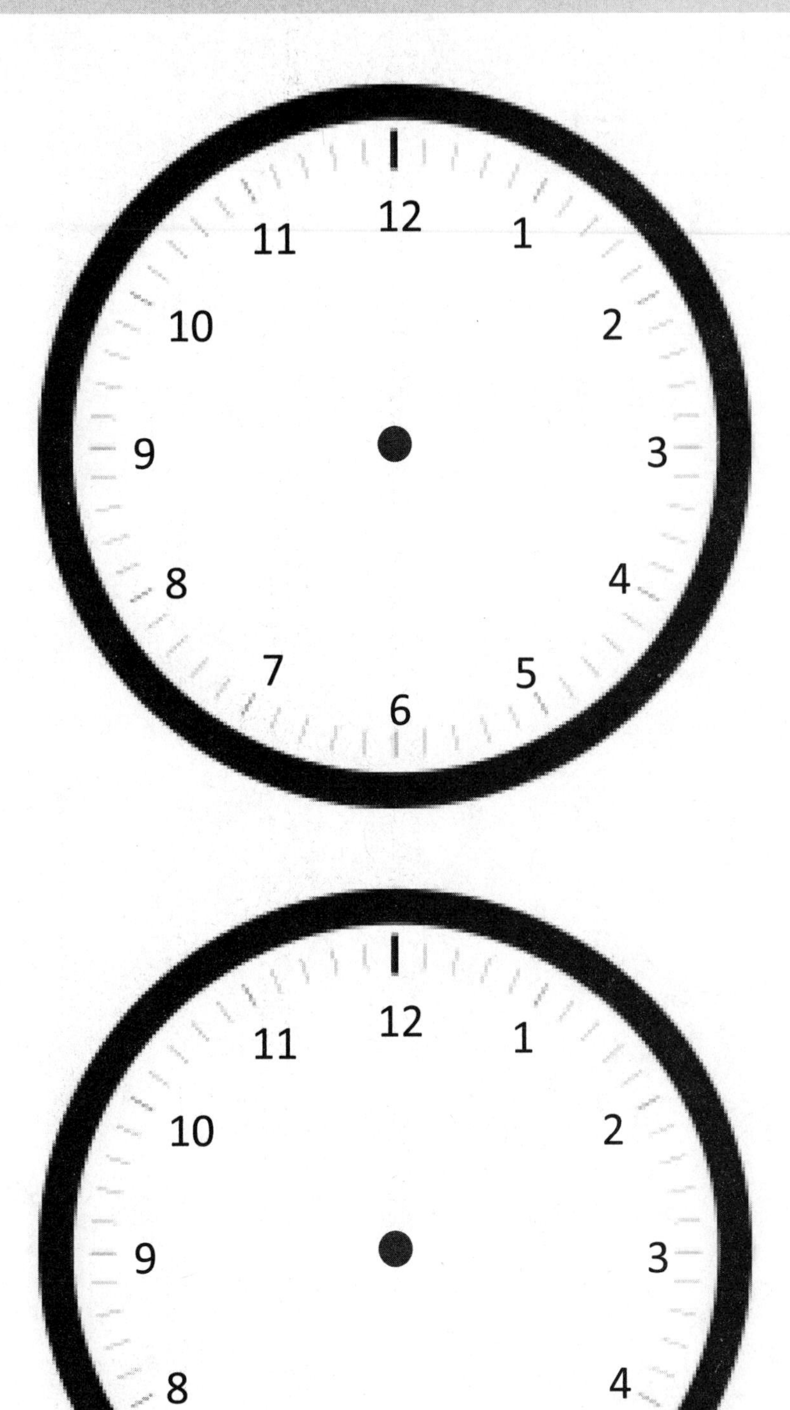

two clocks

©2015 Great Minds. eureka-math.org
G3-M2-SE-B1-1.3.1-01.2016

Name ______________________________ Date ________________

1. Plot a point on the number line for the times shown on the clocks below. Then, draw a line to match the clocks to the points.

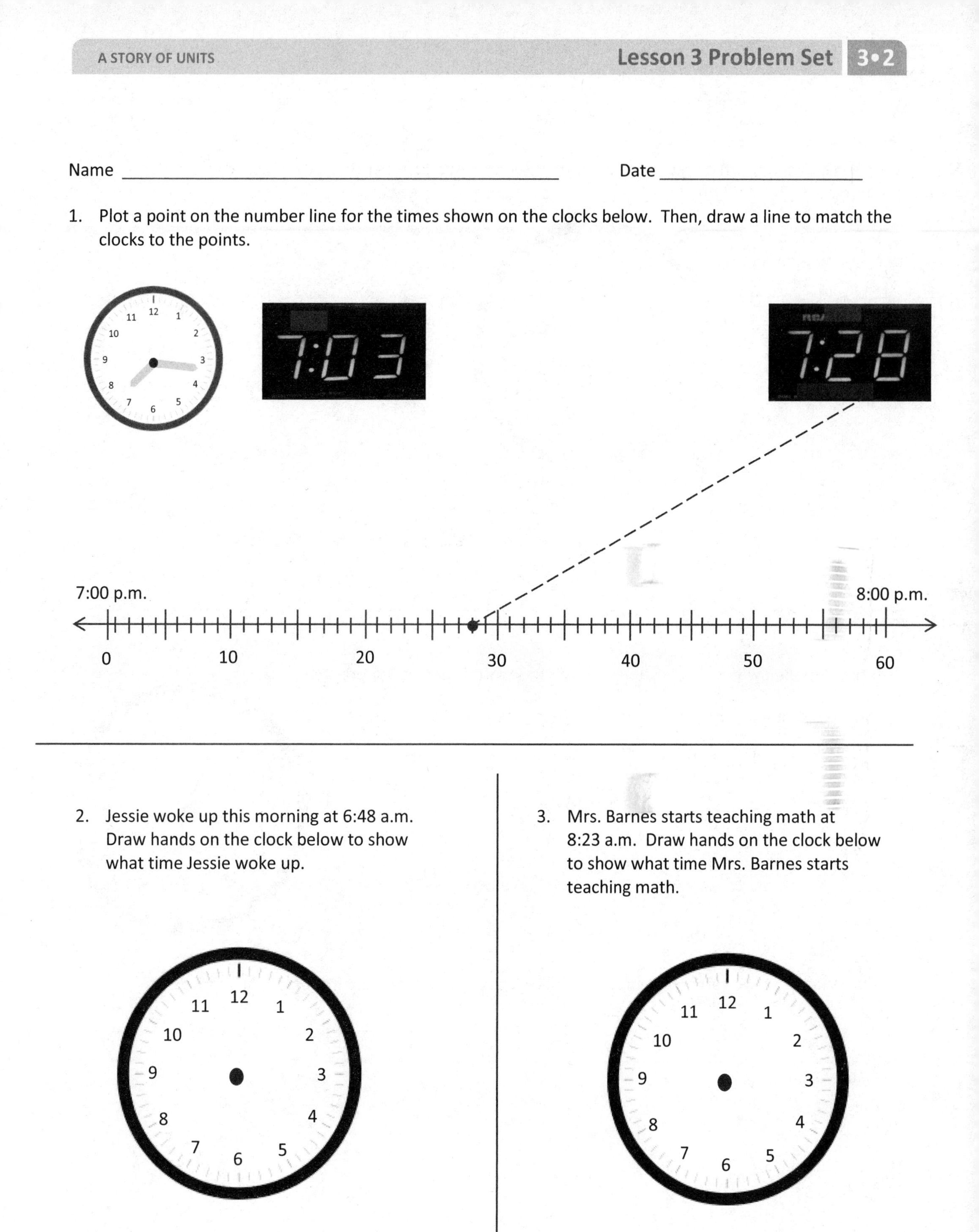

2. Jessie woke up this morning at 6:48 a.m. Draw hands on the clock below to show what time Jessie woke up.

3. Mrs. Barnes starts teaching math at 8:23 a.m. Draw hands on the clock below to show what time Mrs. Barnes starts teaching math.

©2015 Great Minds. eureka-math.org
G3-M2-SE-B1-1.3.1-01.2016

4. The clock shows what time Rebecca finishes her homework. What time does Rebecca finish her homework?

Rebecca finishes her homework at _______________.

---

5. The clock below shows what time Mason's mom drops him off for practice.

   a. What time does Mason's mom drop him off?

   b. Mason's coach arrived 11 minutes before Mason. What time did Mason's coach arrive?

©2015 Great Minds. eureka-math.org
G3-M2-SE-B1-1.3.1-01.2016

Name ______________________________ Date ________________

1. Plot points on the number line for each time shown on a clock below. Then, draw lines to match the clocks to the points.

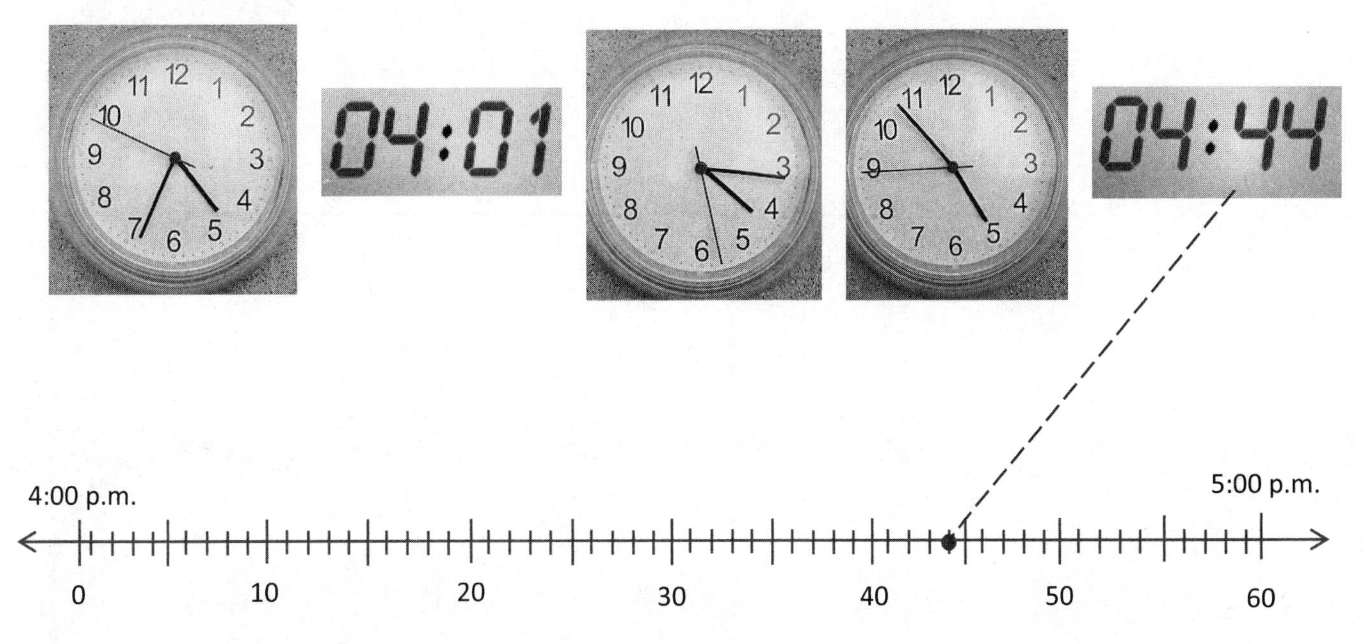

2. Julie eats dinner at 6:07 p.m. Draw hands on the clock below to show what time Julie eats dinner.

3. P.E. starts at 1:32 p.m. Draw hands on the clock below to show what time P.E. starts.

©2015 Great Minds. eureka-math.org
G3-M2-SE-B1-1.3.1-01.2016

4. The clock shows what time Zachary starts playing with his action figures.

   a. What time does he start playing with his action figures?

Start

   b. He plays with his action figures for 23 minutes. What time does he finish playing?

Finish

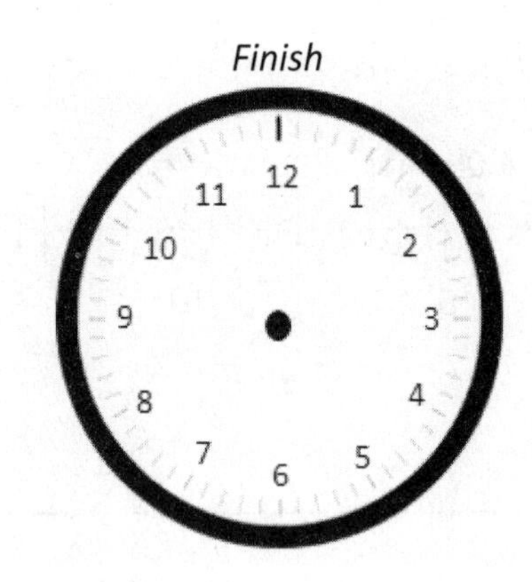

   c. Draw hands on the clock to the right to show what time Zachary finishes playing.

   d. Label the first and last tick marks with 2:00 p.m. and 3:00 p.m. Then, plot Zachary's start and finish times. Label his start time with a *B* and his finish time with an *F*.

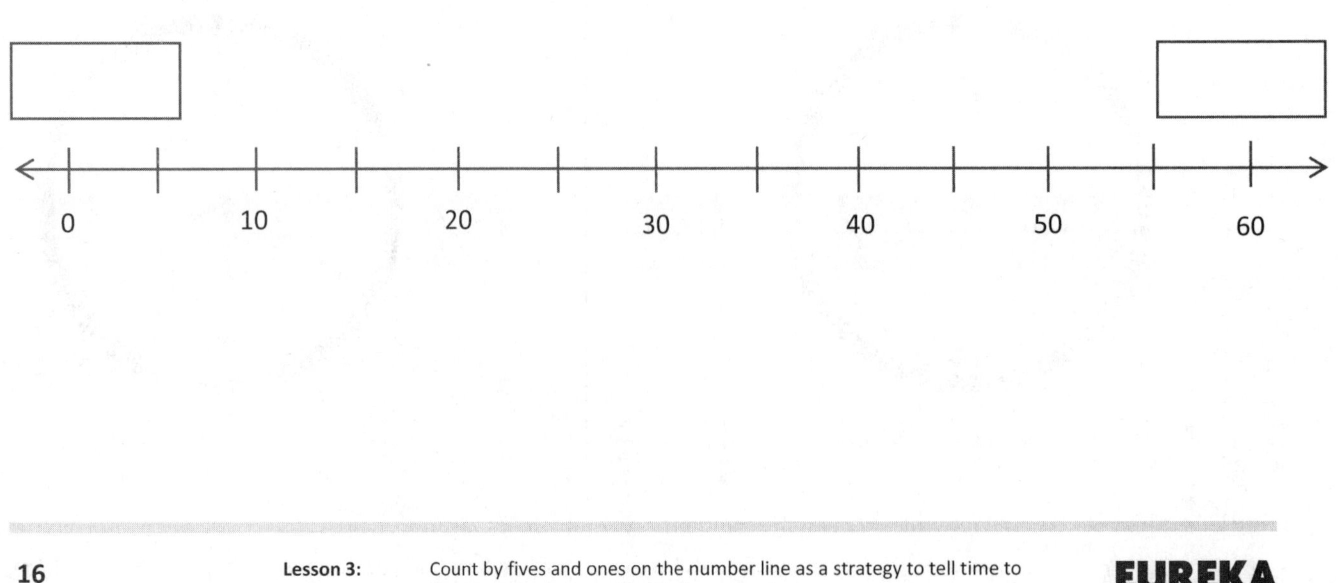

©2015 Great Minds. eureka-math.org
G3-M2-SE-B1-1.3.1-01.2016

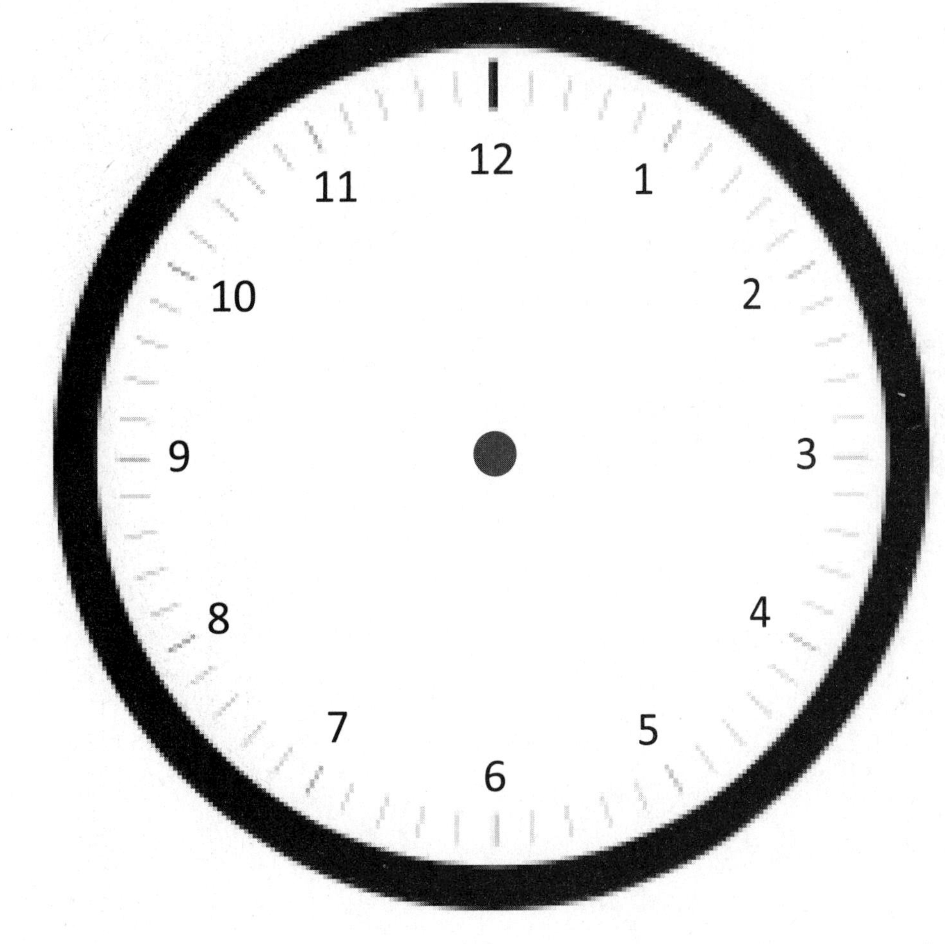

clock

©2015 Great Minds. eureka-math.org
G3-M2-SE-B1-1.3.1-01.2016

Name ______________________________ Date ________________

Use a number line to answer Problems 1 through 5.

1. Cole starts reading at 6:23 p.m. He stops at 6:49 p.m. How many minutes does Cole read?

Cole reads for __________ minutes.

---

2. Natalie finishes piano practice at 2:45 p.m. after practicing for 37 minutes. What time did Natalie's practice start?

Natalie's practice started at __________ p.m.

---

3. Genevieve works on her scrapbook from 11:27 a.m. to 11:58 a.m. How many minutes does she work on her scrapbook?

Genevieve works on her scrapbook for __________ minutes.

---

4. Nate finishes his homework at 4:47 p.m. after working on it for 38 minutes. What time did Nate start his homework?

Nate started his homework at __________ p.m.

---

5. Andrea goes fishing at 9:03 a.m. She fishes for 49 minutes. What time is Andrea done fishing?

Andrea is done fishing at __________ a.m.

©2015 Great Minds. eureka-math.org
G3-M2-SE-B1-1.3.1-01.2016

6. Dion walks to school. The clocks below show when he leaves his house and when he arrives at school. How many minutes does it take Dion to walk to school?

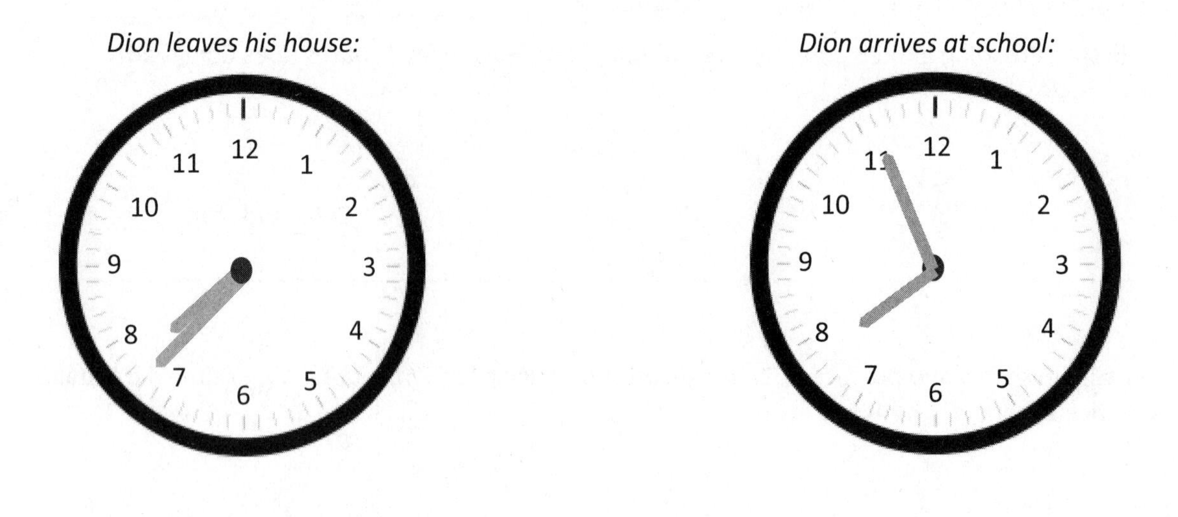

7. Sydney cleans her room for 45 minutes. She starts at 11:13 a.m. What time does Sydney finish cleaning her room?

8. The third-grade chorus performs a musical for the school. The musical lasts 42 minutes. It ends at 1:59 p.m. What time did the musical start?

©2015 Great Minds. eureka-math.org
G3-M2-SE-B1-1.3.1-01.2016

Name ______________________________ Date ______________

Record your homework start time on the clock in Problem 6.

Use a number line to answer Problems 1 through 4.

1. Joy's mom begins walking at 4:12 p.m. She stops at 4:43 p.m. How many minutes does she walk?

Joy's mom walks for _________ minutes.

2. Cassie finishes softball practice at 3:52 p.m. after practicing for 30 minutes. What time did Cassie's practice start?

Cassie's practice started at ___________ p.m.

3. Jordie builds a model from 9:14 a.m. to 9:47 a.m. How many minutes does Jordie spend building his model?

Jordie builds for ___________ minutes.

4. Cara finishes reading at 2:57 p.m. She reads for a total of 46 minutes. What time did Cara start reading?

Cara started reading at ___________ p.m.

©2015 Great Minds. eureka-math.org
G3-M2-SE-B1-1.3.1-01.2016

5. Jenna and her mom take the bus to the mall. The clocks below show when they leave their house and when they arrive at the mall. How many minutes does it take them to get to the mall?

Time when they leave home:

Time when they arrive at the mall:

6. Record your homework start time:

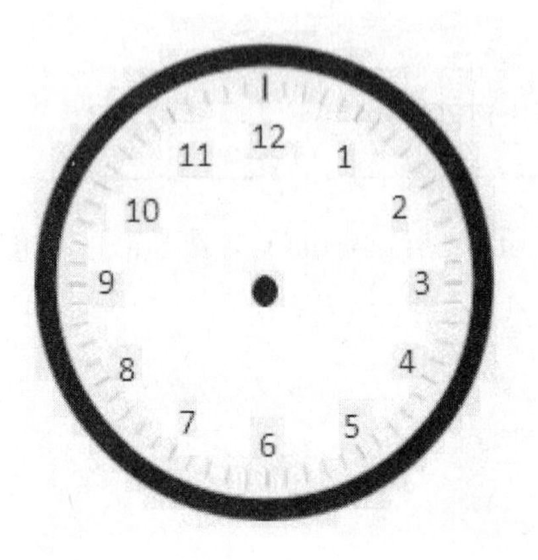

Record the time when you finish Problems 1–5:

How many minutes did you work on Problems 1–5?

EUREKA MATH™

©2015 Great Minds. eureka-math.org
G3-M2-SE-B1-1.3.1-01.2016

number line

©2015 Great Minds. eureka-math.org
G3-M2-SE-B1-1.3.1-01.2016

Name ______________________________ Date ____________________

1. Cole read his book for 25 minutes yesterday and for 28 minutes today. How many minutes did Cole read altogether? Model the problem on the number line, and write an equation to solve.

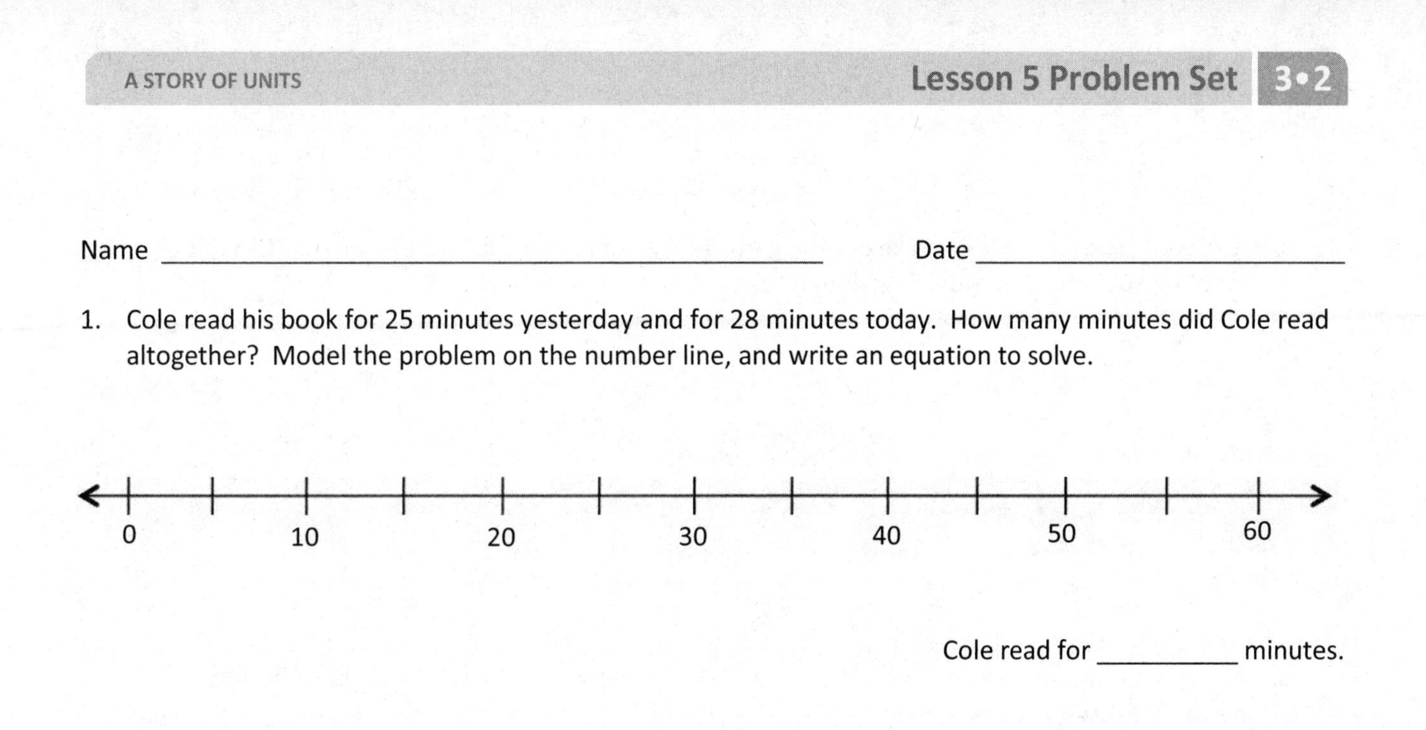

Cole read for ________ minutes.

2. Tessa spends 34 minutes washing her dog. It takes her 12 minutes to shampoo and rinse and the rest of the time to get the dog in the bathtub! How many minutes does Tessa spend getting her dog in the bathtub? Draw a number line to model the problem, and write an equation to solve.

3. Tessa walks her dog for 47 minutes. Jeremiah walks his dog for 30 minutes. How many more minutes does Tessa walk her dog than Jeremiah?

©2015 Great Minds. eureka-math.org
G3-M2-SE-B1-1.3.1-01.2016

4. a. It takes Austin 4 minutes to take out the garbage, 12 minutes to wash the dishes, and 13 minutes to mop the kitchen floor. How long does it take Austin to do his chores?

   b. Austin's bus arrives at 7:55 a.m. If he starts his chores at 7:30 a.m., will he be done in time to meet his bus? Explain your reasoning.

5. Gilberto's cat sleeps in the sun for 23 minutes. It wakes up at the time shown on the clock below. What time did the cat go to sleep?

©2015 Great Minds. eureka-math.org
G3-M2-SE-B1-1.3.1-01.2016

Name ______________________________ Date ________________

1. Abby spent 22 minutes working on her science project yesterday and 34 minutes working on it today. How many minutes did Abby spend working on her science project altogether? Model the problem on the number line, and write an equation to solve.

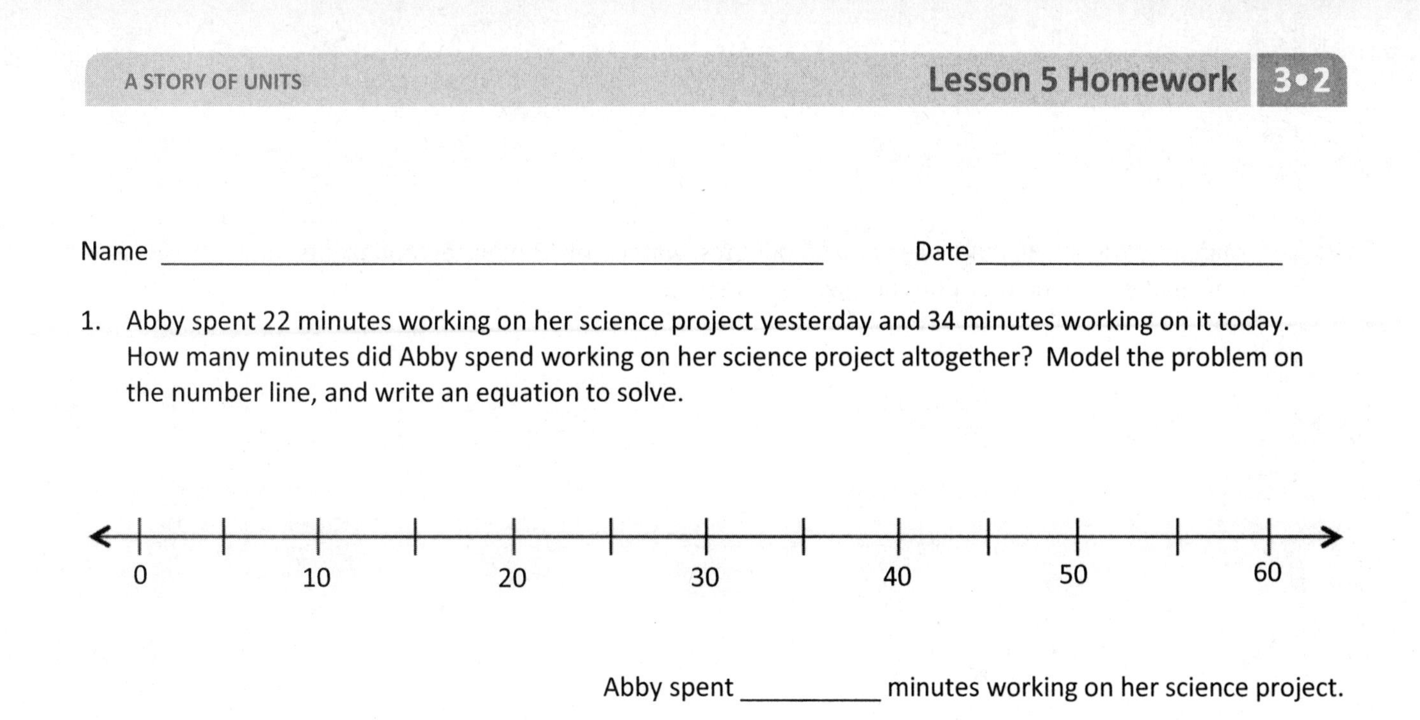

Abby spent ________ minutes working on her science project.

2. Susanna spends a total of 47 minutes working on her project. How many more minutes than Susanna does Abby spend working? Draw a number line to model the problem, and write an equation to solve.

3. Peter practices violin for a total of 55 minutes over the weekend. He practices 25 minutes on Saturday. How many minutes does he practice on Sunday?

©2015 Great Minds. eureka-math.org
G3-M2-SE-B1-1.3.1-01.2016

4. a. Marcus gardens. He pulls weeds for 18 minutes, waters for 13 minutes, and plants for 16 minutes. How many total minutes does he spend gardening?

   b. Marcus wants to watch a movie that starts at 2:55 p.m. It takes 10 minutes to drive to the theater. If Marcus starts the yard work at 2:00 p.m., can he make it on time for the movie? Explain your reasoning.

5. Arelli takes a short nap after school. As she falls asleep, the clock reads 3:03 p.m. She wakes up at the time shown below. How long is Arelli's nap?

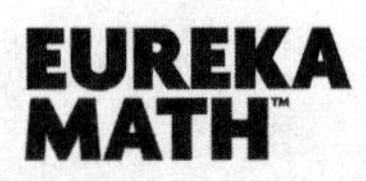

©2015 Great Minds. eureka-math.org
G3-M2-SE-B1-1.3.1-01.2016

Name ______________________________ Date ________________

1. Illustrate and describe the process of making a 1-kilogram weight.

2. Illustrate and describe the process of decomposing 1 kilogram into groups of 100 grams.

3. Illustrate and describe the process of decomposing 100 grams into groups of 10 grams.

©2015 Great Minds. eureka-math.org
G3-M2-SE-B1-1.3.1-01.2016

4. Illustrate and describe the process of decomposing 10 grams into groups of 1 gram.

5. Compare the two place value charts below. How does today's exploration using kilograms and grams relate to your understanding of place value?

1 kilogram	100 grams	10 grams	1 gram

Thousands	Hundreds	Tens	Ones

©2015 Great Minds. eureka-math.org
G3-M2-SE-B1-1.3.1-01.2016

Name ______________________________ Date ________________

1. Use the chart to help you answer the following questions:

1 kilogram	100 grams	10 grams	1 gram

a. Isaiah puts a 10-gram weight on a pan balance. How many 1-gram weights does he need to balance the scale?

b. Next, Isaiah puts a 100-gram weight on a pan balance. How many 10-gram weights does he need to balance the scale?

c. Isaiah then puts a kilogram weight on a pan balance. How many 100-gram weights does he need to balance the scale?

d. What pattern do you notice in Parts (a–c)?

©2015 Great Minds. eureka-math.org
G3-M2-SE-B1-1.3.1-01.2016

2. Read each digital scale. Write each weight using the word *kilogram* or *gram* for each measurement.

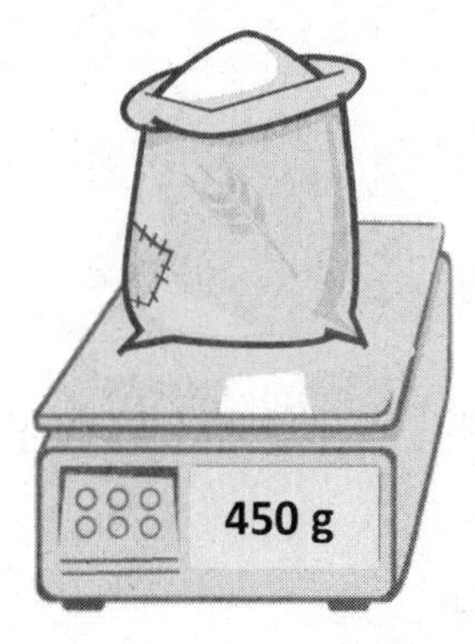

________________ ________________ ________________

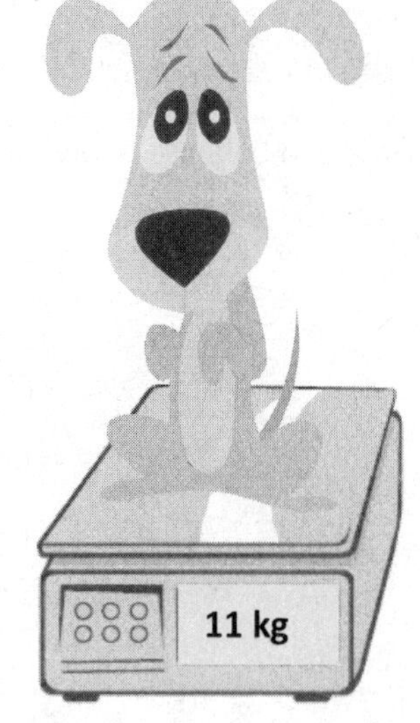

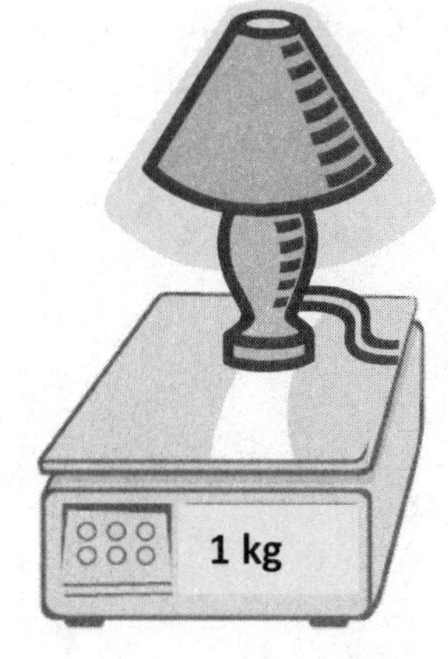

________________ ________________ ________________

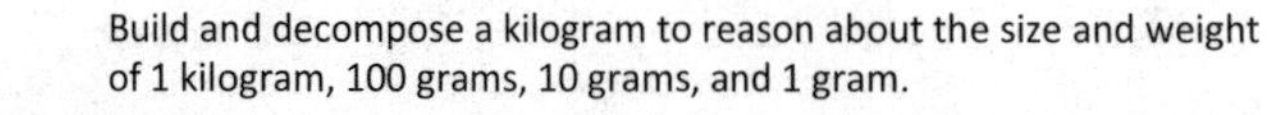

©2015 Great Minds. eureka math.org
G3-M2-SE-B1-1.3.1-01.2016

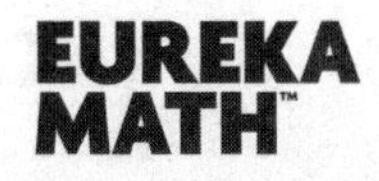

Name ______________________________ Date ________________

Work with a partner. Use the corresponding weights to estimate the weight of objects in the classroom. Then, check your estimate by weighing on a scale.

A.

Objects that Weigh About **1 Kilogram**	Actual Weight

B.

Objects that Weigh About **100 Grams**	Actual Weight

C.

Objects that Weigh About **10 Grams**	Actual Weight

D.

Objects that Weigh About **1 Gram**	Actual Weight

©2015 Great Minds. eureka-math.org
G3-M2-SE-B1-1.3.1-01.2016

E. Circle the correct unit of weight for each estimation.

1. A box of cereal weighs about 350 (grams / kilograms).

2. A watermelon weighs about 3 (grams / kilograms).

3. A postcard weighs about 6 (grams / kilograms).

4. A cat weighs about 4 (grams / kilograms).

5. A bicycle weighs about 15 (grams / kilograms).

6. A lemon weighs about 58 (grams / kilograms).

F. During the exploration, Derrick finds that his bottle of water weighs the same as a 1-kilogram bag of rice. He then exclaims, "Our class laptop weighs the same as 2 bottles of water!" How much does the laptop weigh in kilograms? Explain your reasoning.

G. Nessa tells her brother that 1 kilogram of rice weighs the same as 10 bags containing 100 grams of beans each. Do you agree with her? Explain why or why not.

©2015 Great Minds. eureka-math.org
G3-M2-SE-B1-1.3.1-01.2016

Name ______________________________________ Date ____________________

1. Match each object with its approximate weight.

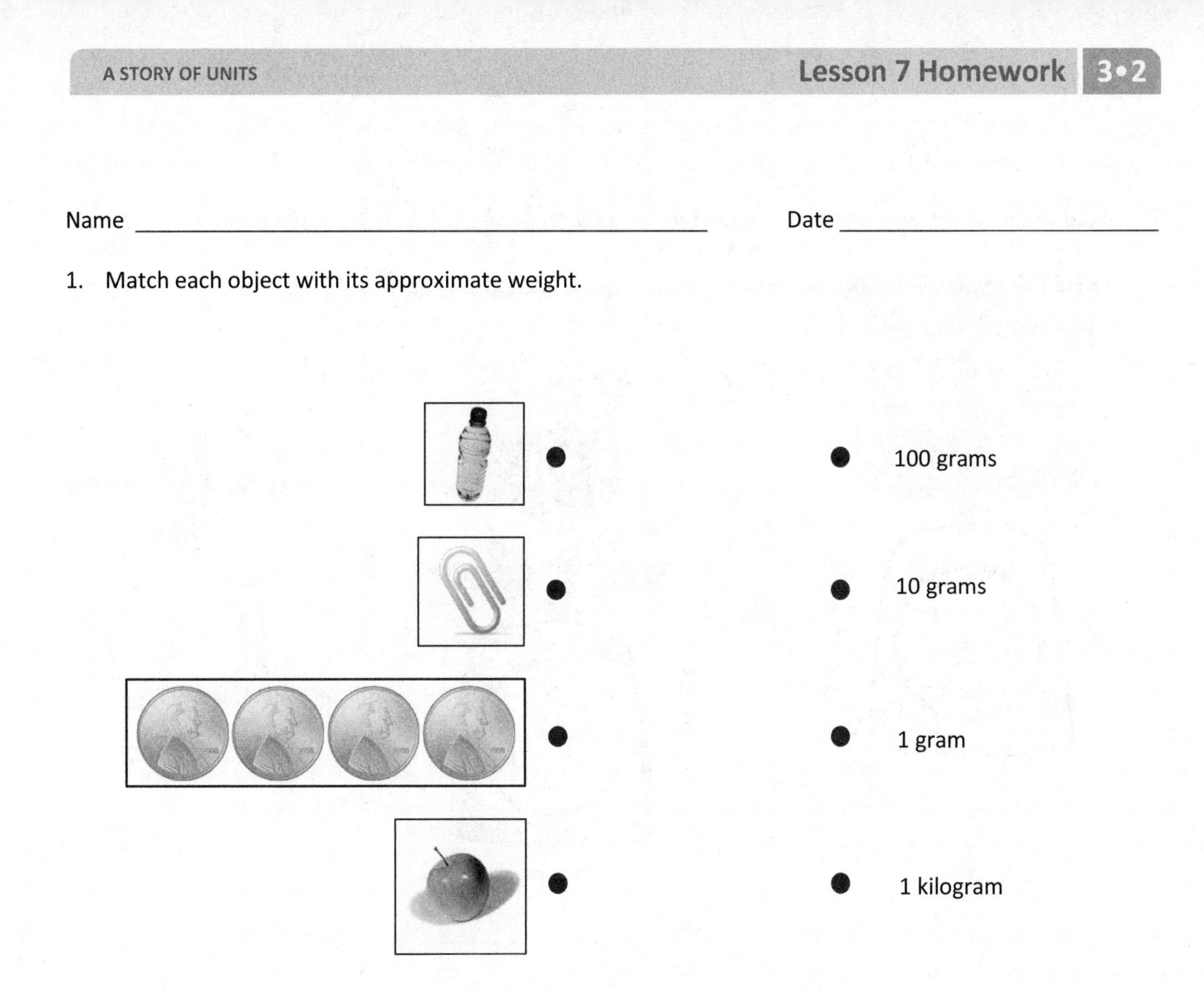

2. Alicia and Jeremy weigh a cell phone on a digital scale. They write down 113 but forget to record the unit. Which unit of measurement is correct, grams or kilograms? How do you know?

©2015 Great Minds. eureka-math.org
G3-M2-SE-B1-1.3.1-01.2016

3. Read and write the weights below. Write the word *kilogram* or *gram* with the measurement.

________________ ________________ ________________

________________ ________________

©2015 Great Minds. eureka-math.org
G3-M2-SE-B1-1.3.1-01.2016

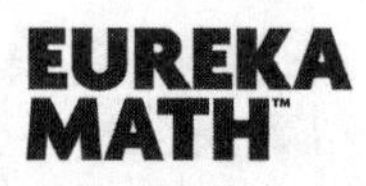

Name ______________________________ Date ____________________

1. Tim goes to the market to buy fruits and vegetables. He weighs some string beans and some grapes.

List the weights for both the string beans and grapes.

The string beans weigh __________ grams.

The grapes weigh __________ grams.

---

2. Use tape diagrams to model the following problems. Keiko and her brother Jiro get weighed at the doctor's office. Keiko weighs 35 kilograms, and Jiro weighs 43 kilograms.

   a. What is Keiko and Jiro's total weight?

   Keiko and Jiro weigh ________ kilograms.

   b. How much heavier is Jiro than Keiko?

   Jiro is ________ kilograms heavier than Keiko.

©2015 Great Minds. eureka-math.org
G3-M2-SE-B1-1.3.1-01.2016

3. Jared estimates that his houseplant is as heavy as a 5-kilogram bowling ball. Draw a tape diagram to estimate the weight of 3 houseplants.

---

4. Jane and her 8 friends go apple picking. They share what they pick equally. The total weight of the apples they pick is shown to the right.

   a. About how many kilograms of apples will Jane take home?

   b. Jane estimates that a pumpkin weighs about as much as her share of the apples. About how much do 7 pumpkins weigh altogether?

©2015 Great Minds. eureka-math.org
G3-M2-SE-B1-1.3.1-01.2016

Name ______________________________ Date ____________________

1. The weights of 3 fruit baskets are shown below.

Basket A	Basket B	Basket C
12 kg	8 kg	16 kg

a. Basket ______ is the heaviest.

b. Basket ______ is the lightest.

c. Basket A is __________ kilograms heavier than Basket B.

d. What is the total weight of all three baskets?

2. Each journal weighs about 280 grams. What is total weight of 3 journals?

3. Ms. Rios buys 453 grams of strawberries. She has 23 grams left after making smoothies. How many grams of strawberries did she use?

©2015 Great Minds. eureka-math.org
G3-M2-SE-B1-1.3.1-01.2016

4. Andrea's dad is 57 kilograms heavier than Andrea. Andrea weighs 34 kilograms.

   a. How much does Andrea's dad weigh?

   b. How much do Andrea and her dad weigh in total?

5. Jennifer's grandmother buys carrots at the farm stand. She and her 3 grandchildren equally share the carrots. The total weight of the carrots she buys is shown below.

   a. How many kilograms of carrots will Jennifer get?

   b. Jennifer uses 2 kilograms of carrots to bake muffins. How many kilograms of carrots does she have left?

©2015 Great Minds. eureka-math.org
G3-M2-SE-B1-1.3.1-01.2016

Name ____________________________________ Date ________________

Part 1

a. Predict whether each container holds less than, more than, or about the same as 1 liter.

Container 1 holds	less than / more than / about the same as	1 liter.	Actual:
Container 2 holds	less than / more than / about the same as	1 liter.	Actual:
Container 3 holds	less than / more than / about the same as	1 liter.	Actual:
Container 4 holds	less than / more than / about the same as	1 liter.	Actual:

b. After measuring, what surprised you? Why?

Part 2

c. Illustrate and describe the process of decomposing 1 liter of water into 10 smaller units.

©2015 Great Minds. eureka-math.org
G3-M2-SE-B1-1.3.1-01.2016

d. Illustrate and describe the process of decomposing Cup K into 10 smaller units.

e. Illustrate and describe the process of decomposing Cup L into 10 smaller units.

f. What is the same about decomposing 1 liter into milliliters and decomposing 1 kilogram into grams?

g. One liter of water weighs 1 kilogram. How much does 1 milliliter of water weigh? Explain how you know.

©2015 Great Minds. eureka-math.org
G3-M2-SE-B1-1.3.1-01.2016

Name ______________________________ Date ________________

1. Find containers at home that have a capacity of about 1 liter. Use the labels on containers to help you identify them.

   a.

Name of Container
Example: Carton of orange juice

   b. Sketch the containers. How do their sizes and shapes compare?

2. The doctor prescribes Mrs. Larson 5 milliliters of medicine each day for 3 days. How many milliliters of medicine will she take altogether?

©2015 Great Minds. eureka-math.org
G3-M2-SE-B1-1.3.1-01.2016

3. Mrs. Goldstein pours 3 juice boxes into a bowl to make punch. Each juice box holds 236 milliliters. How much juice does Mrs. Goldstein pour into the bowl?

---

4. Daniel's fish tank holds 24 liters of water. He uses a 4-liter bucket to fill the tank. How many buckets of water are needed to fill the tank?

---

5. Sheila buys 15 liters of paint to paint her house. She pours the paint equally into 3 buckets. How many liters of paint are in each bucket?

©2015 Great Minds. eureka-math.org
G3-M2-SE-B1-1.3.1-01.2016

Name_______________________________________ Date ___________________

1. Label the vertical number line on the container to the right. Answer the questions below.

   a. What did you label as the halfway mark? Why?

   b. Explain how pouring each plastic cup of water helped you create a vertical number line.

   c. If you pour out 300 mL of water, how many mL are left in the container?

100 mL

2. How much liquid is in each container?

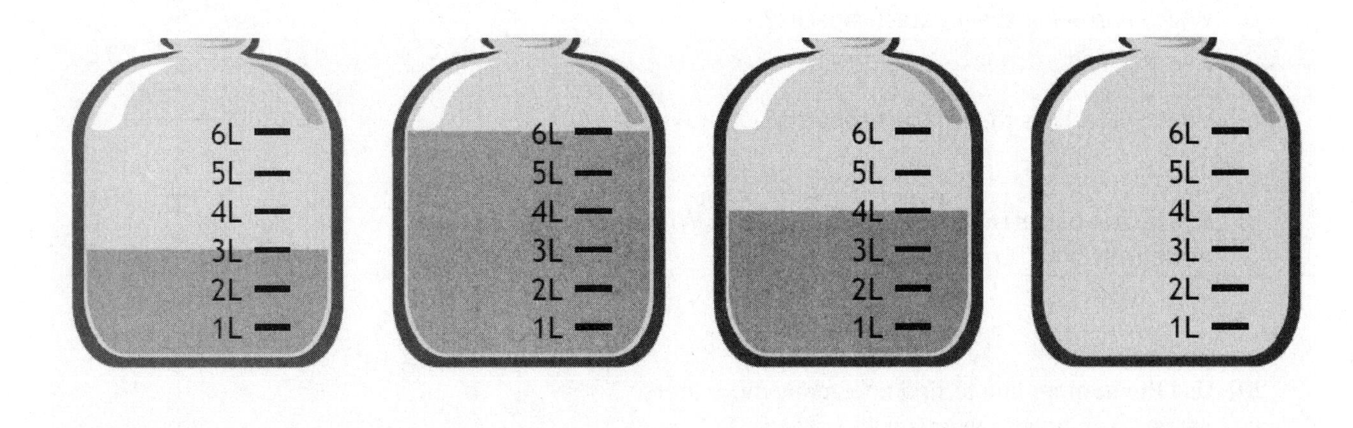

_____________ _____________ _____________ _____________

©2015 Great Minds. eureka-math.org
G3-M2-SE-B1-1.3.1-01.2016

3. Estimate the amount of liquid in each container to the nearest hundred milliliters.

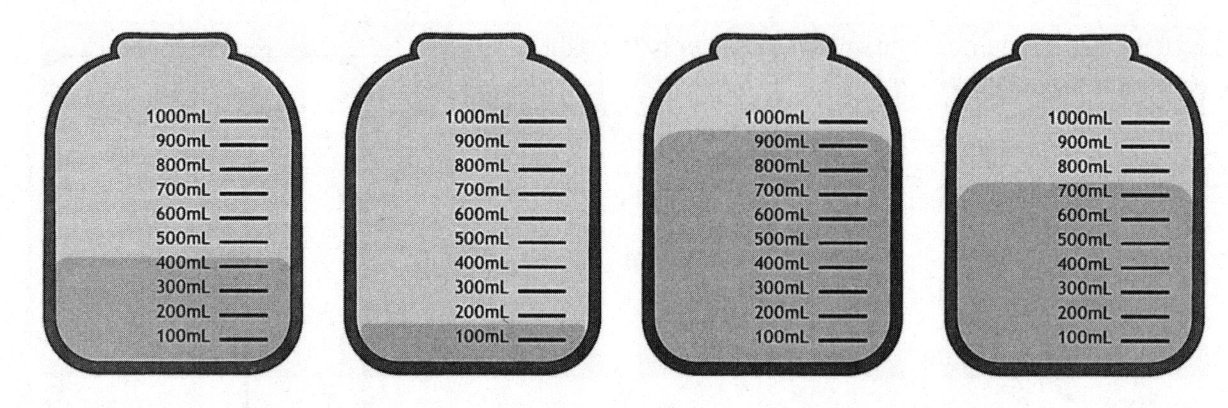

____________ ____________ ____________ ____________

4. The chart below shows the capacity of 4 barrels.

Barrel A	75 liters
Barrel B	68 liters
Barrel C	96 liters
Barrel D	52 liters

100 L
90 L
80 L
Barrel A ***75 L***
70 L
60 L
50 L
40 L

a. Label the number line to show the capacity of each barrel. Barrel A has been done for you.

b. Which barrel has the greatest capacity?

c. Which barrel has the smallest capacity?

d. Ben buys a barrel that holds about 70 liters. Which barrel did he most likely buy? Explain why.

e. Use the number line to find how many more liters Barrel C can hold than Barrel B.

©2015 Great Minds. eureka-math.org
G3-M2-SE-B1-1.3.1-01.2016

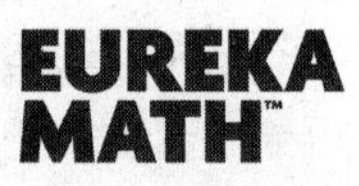

Name____________________________________ Date __________________

1. How much liquid is in each container?

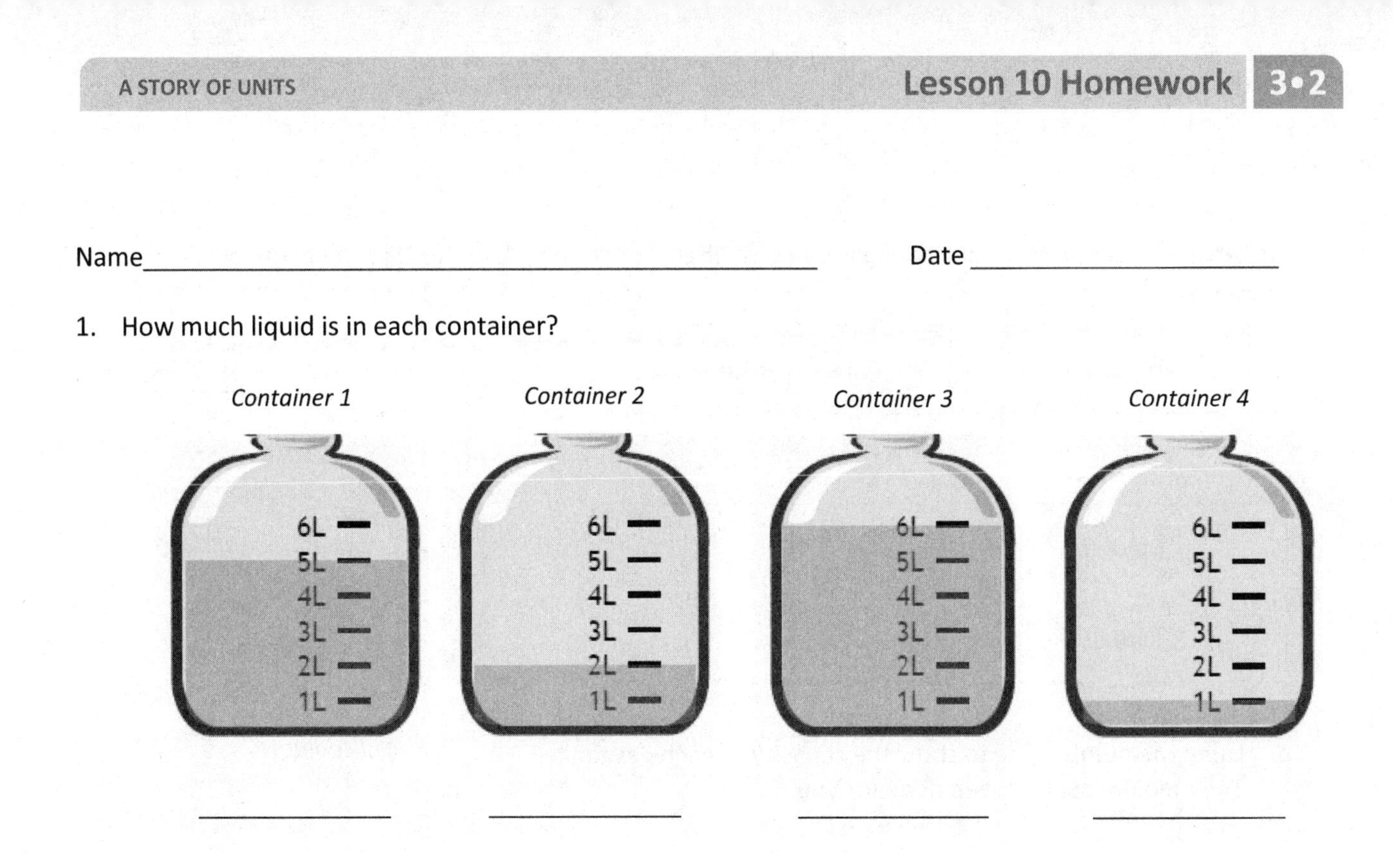

2. Jon pours the contents of Container 1 and Container 3 above into an empty bucket. How much liquid is in the bucket after he pours the liquid?

3. Estimate the amount of liquid in each container to the nearest liter.

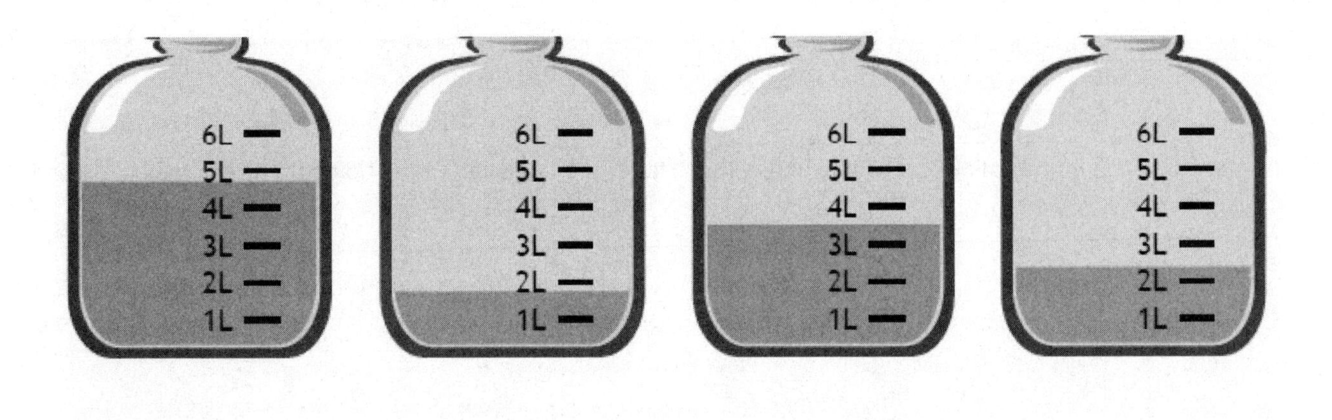

©2015 Great Minds. eureka-math.org
G3-M2-SE-B1-1.3.1-01.2016

4. Kristen is comparing the capacity of gas tanks in different size cars. Use the chart below to answer the questions.

Size of Car	Capacity in Liters
Large	74
Medium	57
Small	42

80 L
70 L
60 L
Medium
50 L
40 L
30 L
20 L

a. Label the number line to show the capacity of each gas tank. The medium car has been done for you.

b. Which car's gas tank has the greatest capacity?

c. Which car's gas tank has the smallest capacity?

d. Kristen's car has a gas tank capacity of about 60 liters. Which car from the chart has about the same capacity as Kristen's car?

e. Use the number line to find how many more liters the large car's tank holds than the small car's tank.

©2015 Great Minds. eureka-math.org
G3-M2-SE-B1-1.3.1-01.2016

Name ______________________________ Date ______________

1. The total weight in grams of a can of tomatoes and a jar of baby food is shown to the right.

   a. The jar of baby food weighs 113 grams. How much does the can of tomatoes weigh?

   b. How much more does the can of tomatoes weigh than the jar of baby food?

671 g

2. The weight of a pen in grams is shown to the right.

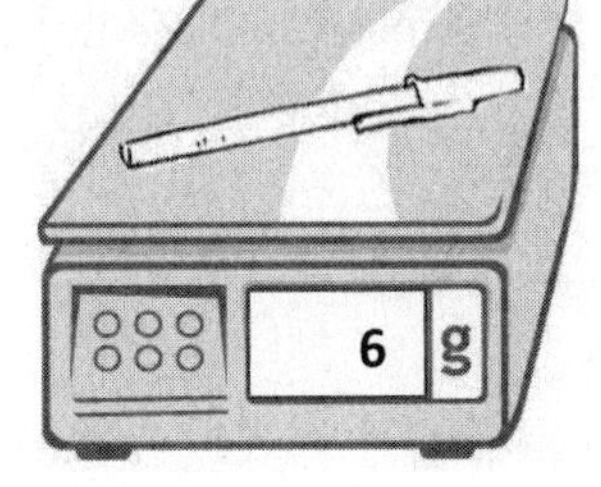

   a. What is the total weight of 10 pens?

   b. An empty box weighs 82 grams. What is the total weight of a box of 10 pens?

3. The total weight of an apple, lemon, and banana in grams is shown to the right.

   a. If the apple and lemon together weigh 317 grams, what is the weight of the banana?

   b. If we know the lemon weighs 68 grams less than the banana, how much does the lemon weigh?

   c. What is the weight of the apple?

©2015 Great Minds. eureka-math.org
G3-M2-SE-B1-1.3.1-01.2016

4. A frozen turkey weighs about 5 kilograms. The chef orders 45 kilograms of turkey. Use a tape diagram to find about how many frozen turkeys he orders.

---

5. A recipe requires 300 milliliters of milk. Sara decides to triple the recipe for dinner. How many milliliters of milk does she need to cook dinner?

---

6. Marian pours a full container of water equally into buckets. Each bucket has a capacity of 4 liters. After filling 3 buckets, she still has 2 liters left in her container. What is the capacity of her container?

©2015 Great Minds. eureka-math.org
G3-M2-SE-B1-1.3.1-01.2016

Name ______________________________ Date ______________

1. Karina goes on a hike. She brings a notebook, a pencil, and a camera. The weight of each item is shown in the chart. What is the total weight of all three items?

Item	Weight
Notebook	312 g
Pencil	10 g
Camera	365 g

The total weight is __________ grams.

---

2. Together a horse and its rider weigh 729 kilograms. The horse weighs 625 kilograms. How much does the rider weigh?

The rider weighs __________ kilograms.

©2015 Great Minds. eureka-math.org
G3-M2-SE-B1-1.3.1-01.2016

3. Theresa's soccer team fills up 6 water coolers before the game. Each water cooler holds 9 liters of water. How many liters of water do they fill?

---

4. Dwight purchased 48 kilograms of fertilizer for his vegetable garden. He needs 6 kilograms of fertilizer for each bed of vegetables. How many beds of vegetables can he fertilize?

---

5. Nancy bakes 7 cakes for the school bake sale. Each cake requires 5 milliliters of oil. How many milliliters of oil does she use?

©2015 Great Minds. eureka-math.org
G3-M2-SE-B1-1.3.1-01.2016

Name ______________________________ Date ________________

1. Work with a partner. Use a ruler or a meter stick to complete the chart below.

Object	Measurement (in cm)	The object measures between (which two tens)...	Length rounded to the nearest 10 cm
Example: My shoe	23 cm	20 and 30 cm	20 cm
Long side of a desk		______ and ______ cm	
A new pencil		______ and ______ cm	
Short side of a piece of paper		______ and ______ cm	
Long side of a piece of paper		______ and ______ cm	

2. Work with a partner. Use a digital scale to complete the chart below.

Bag	Measurement (in g)	The bag of rice measures between (which two tens)...	Weight rounded to the nearest 10 g
Example: Bag A	8 g	0 and 10 g	10 g
Bag B		______ and ______ g	
Bag C		______ and ______ g	
Bag D		______ and ______ g	
Bag E		______ and ______ g	

©2015 Great Minds. eureka-math.org
G3-M2-SE-B1-1.3.1-01.2016

3. Work with a partner. Use a beaker to complete the chart below.

Container	Measurement (in mL)	The container measures between (which two tens)...	Liquid volume rounded to the nearest 10 mL
Example: Container A	33 mL	30 and 40 mL	30 mL
Container B		______ and ______ mL	
Container C		______ and ______ mL	
Container D		______ and ______ mL	
Container E		______ and ______ mL	

4. Work with a partner. Use a clock to complete the chart below.

Activity	Actual time	The activity measures between (which two tens)...	Time rounded to the nearest 10 minutes
Example: Time we started math	10:03	10:00 and 10:10	10:00
Time I started the Problem Set		______ and ______	
Time I finished Station 1		______ and ______	
Time I finished Station 2		______ and ______	
Time I finished Station 3		______ and ______	

©2015 Great Minds. eureka-math.org
G3-M2-SE-B1-1.3.1-01.2016

Name ________________________________ Date ________________

1. Complete the chart. Choose objects, and use a ruler or meter stick to complete the last two on your own.

Object	Measurement (in cm)	The object measures between (which two tens)...	Length rounded to the nearest 10 cm
Length of desk	66 cm	______ and ______ cm	
Width of desk	48 cm	______ and ______ cm	
Width of door	81 cm	______ and ______ cm	
		______ and ______ cm	
		______ and ______ cm	

2. Gym class ends at 10:27 a.m. Round the time to the nearest 10 minutes.

Gym class ends at about ______ a.m.

3. Measure the liquid in the beaker to the nearest 10 milliliters.

60 mL
50 mL
40 mL
30 mL
20 mL
10 mL

There are about ________ milliliters in the beaker.

Lesson 12: Round two-digit measurements to the nearest ten on the vertical number line.

55

©2015 Great Minds. eureka-math.org
G3-M2-SE-B1-1.3.1-01.2016

4. Mrs. Santos' weight is shown on the scale. Round her weight to the nearest 10 kilograms.

Mrs. Santos' weight is _________ kilograms.

Mrs. Santos weighs about _________ kilograms.

5. A zookeeper weighs a chimp. Round the chimp's weight to the nearest 10 kilograms.

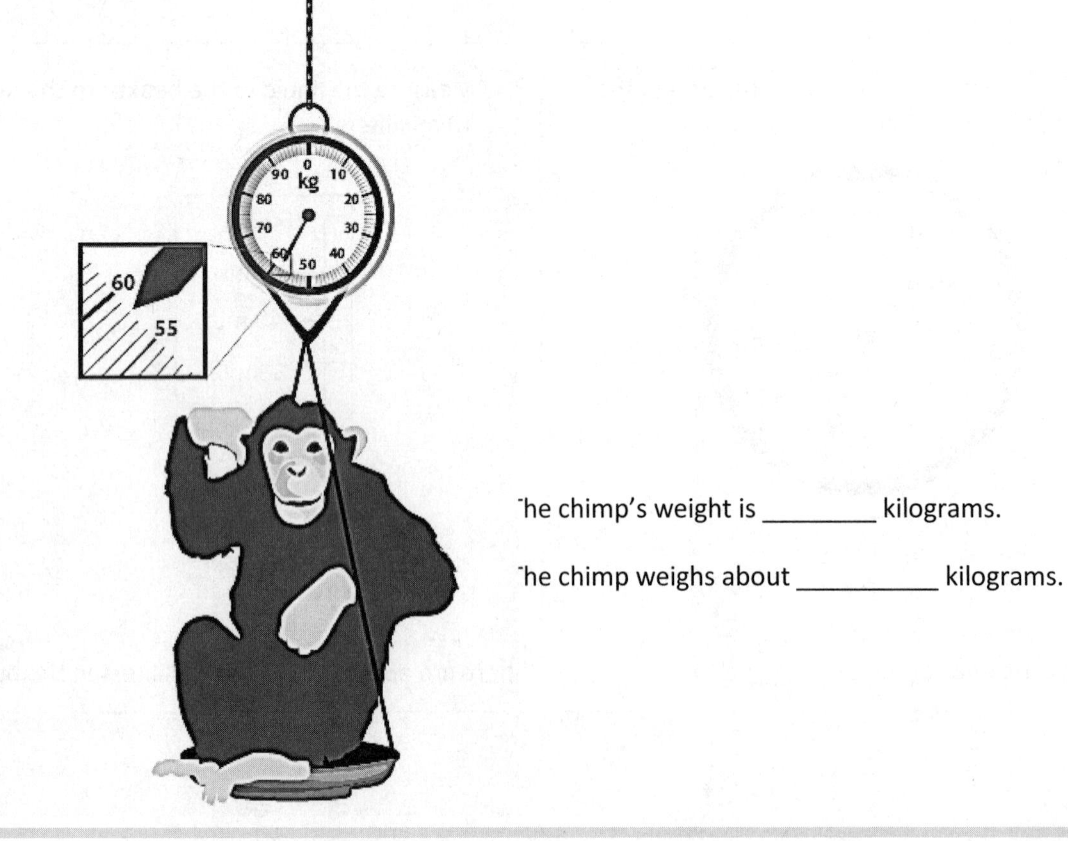

ˉhe chimp's weight is ________ kilograms.

ˉhe chimp weighs about __________ kilograms.

©2015 Great Minds. eureka-math.org
G3-M2-SE-B1-1.3.1-01.2016

Name ______________________________ Date ________________

1. Round to the nearest ten. Use the number line to model your thinking.

a. 32 ≈ _________

b. 36 ≈ _________

c. 62 ≈ _________

d. 162 ≈ _________

e. 278 ≈ _________

f. 405 ≈ _________

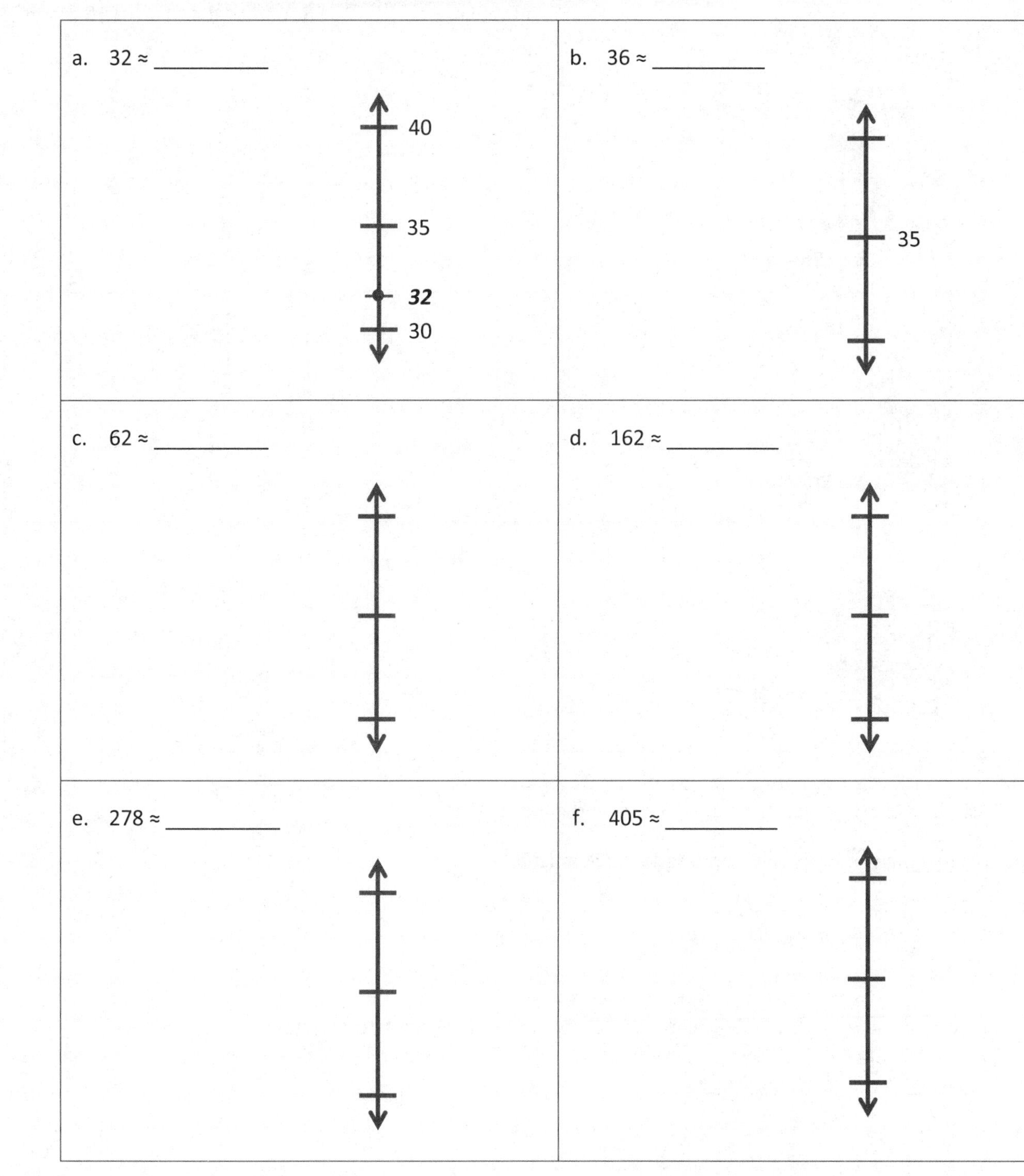

©2015 Great Minds. eureka-math.org
G3-M2-SE-B1-1.3.1-01.2016

2. Round the weight of each item to the nearest 10 grams. Draw number lines to model your thinking.

Item	Number Line	Round to the nearest 10 grams
CHIPS 36 grams		
52 grams		
142 grams		

3. Carl's basketball game begins at 3:03 p.m. and ends at 3:51 p.m.

a. How many minutes did Carl's basketball game last?

b. Round the total number of minutes in the game to the nearest 10 minutes.

©2015 Great Minds. eureka-math.org
G3-M2-SE-B1-1.3.1-01.2016

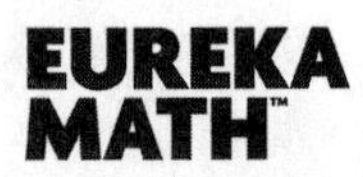

Name ______________________________ Date ______________

1. Round to the nearest ten. Use the number line to model your thinking.

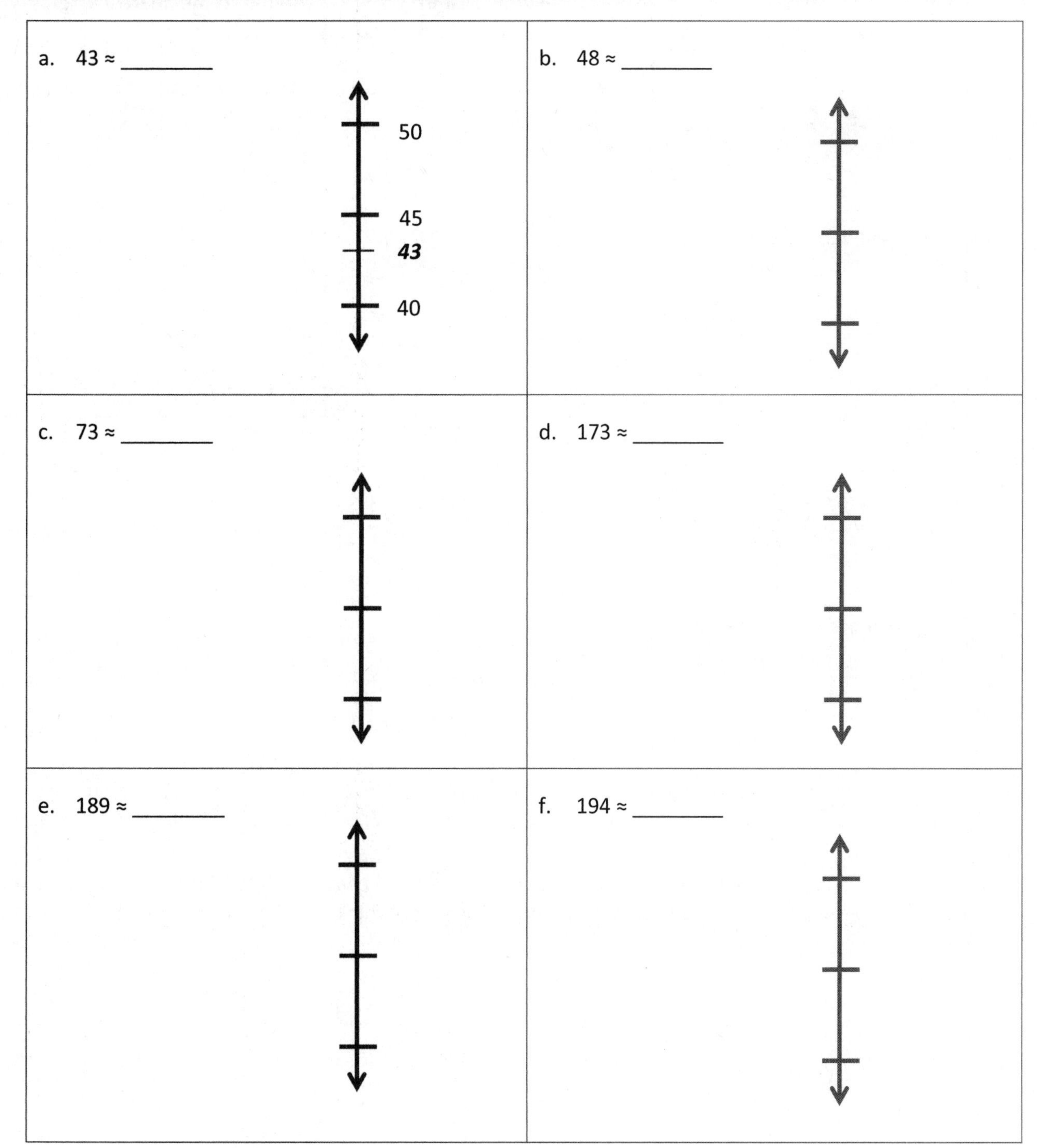

a. 43 ≈ _____

50

45

43

40

b. 48 ≈ _____

c. 73 ≈ _____

d. 173 ≈ _____

e. 189 ≈ _____

f. 194 ≈ _____

EUREKA
MATH™

Lesson 13: Round two- and three-digit numbers to the nearest ten on the vertical
number line.

59

©2015 Great Minds. eureka-math.org
G3-M2-SE-B1-1.3.1-01.2016

2. Round the weight of each item to the nearest 10 grams. Draw number lines to model your thinking.

Item	Number Line	Round to the nearest 10 grams
Cereal bar: 45 grams		
Loaf of bread: 673 grams		

3. The Garden Club plants rows of carrots in the garden. One seed packet weighs 28 grams. Round the total weight of 2 seed packets to the nearest 10 grams. Model your thinking using a number line.

©2015 Great Minds. eureka-math.org
G3-M2-SE-B1-1.3.1-01.2016

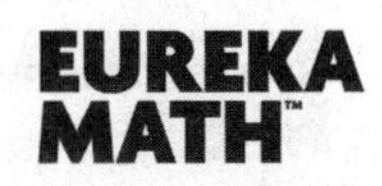

Name ______________________________ Date ______________

1. Round to the nearest hundred. Use the number line to model your thinking.

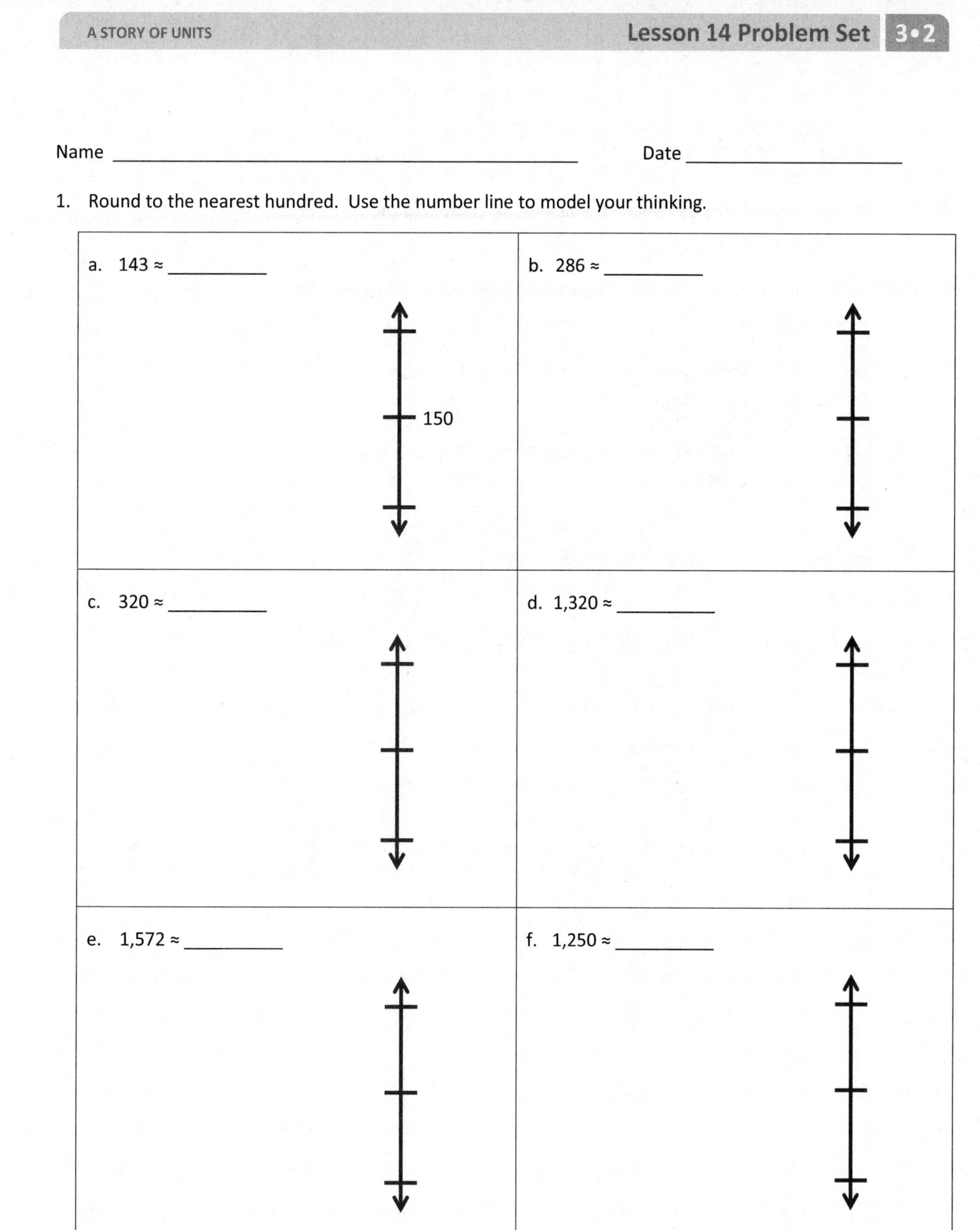

©2015 Great Minds. eureka-math.org
G3-M2-SE-B1-1.3.1-01.2016

2. Complete the chart.

a. Shauna has 480 stickers. Round the number of stickers to the nearest hundred.	
b. There are 525 pages in a book. Round the number of pages to the nearest hundred.	
c. A container holds 750 milliliters of water. Round the capacity to the nearest 100 milliliters.	
d. Glen spends $1,297 on a new computer. Round the amount Glen spends to the nearest $100.	
e. The drive between two cities is 1,842 kilometers. Round the distance to the nearest 100 kilometers.	

3. Circle the numbers that round to 600 when rounding to the nearest hundred.

527 550 639 681 713 603

4. The teacher asks students to round 1,865 to the nearest hundred. Christian says that it is one thousand, nine hundred. Alexis disagrees and says it is 19 hundreds. Who is correct? Explain your thinking.

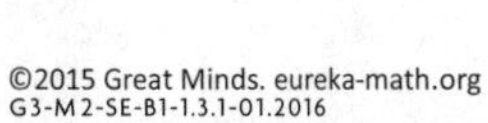
©2015 Great Minds. eureka-math.org
G3-M2-SE-B1-1.3.1-01.2016

Name ______________________________ Date ______________

1. Round to the nearest hundred. Use the number line to model your thinking.

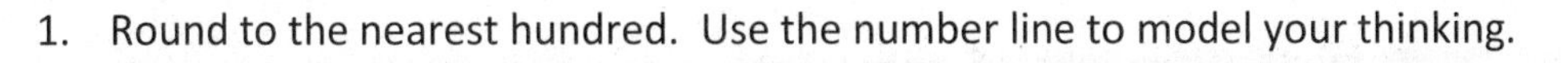

a. 156 ≈ _____

b. 342 ≈ _____

c. 260 ≈ _____

d. 1,260 ≈ _____

e. 1,685 ≈ _____

f. 1,804 ≈ _____

Lesson 14: Round to the nearest hundred on the vertical number line.

63

Lesson 14: Round to the nearest hundred on the vertical number line.

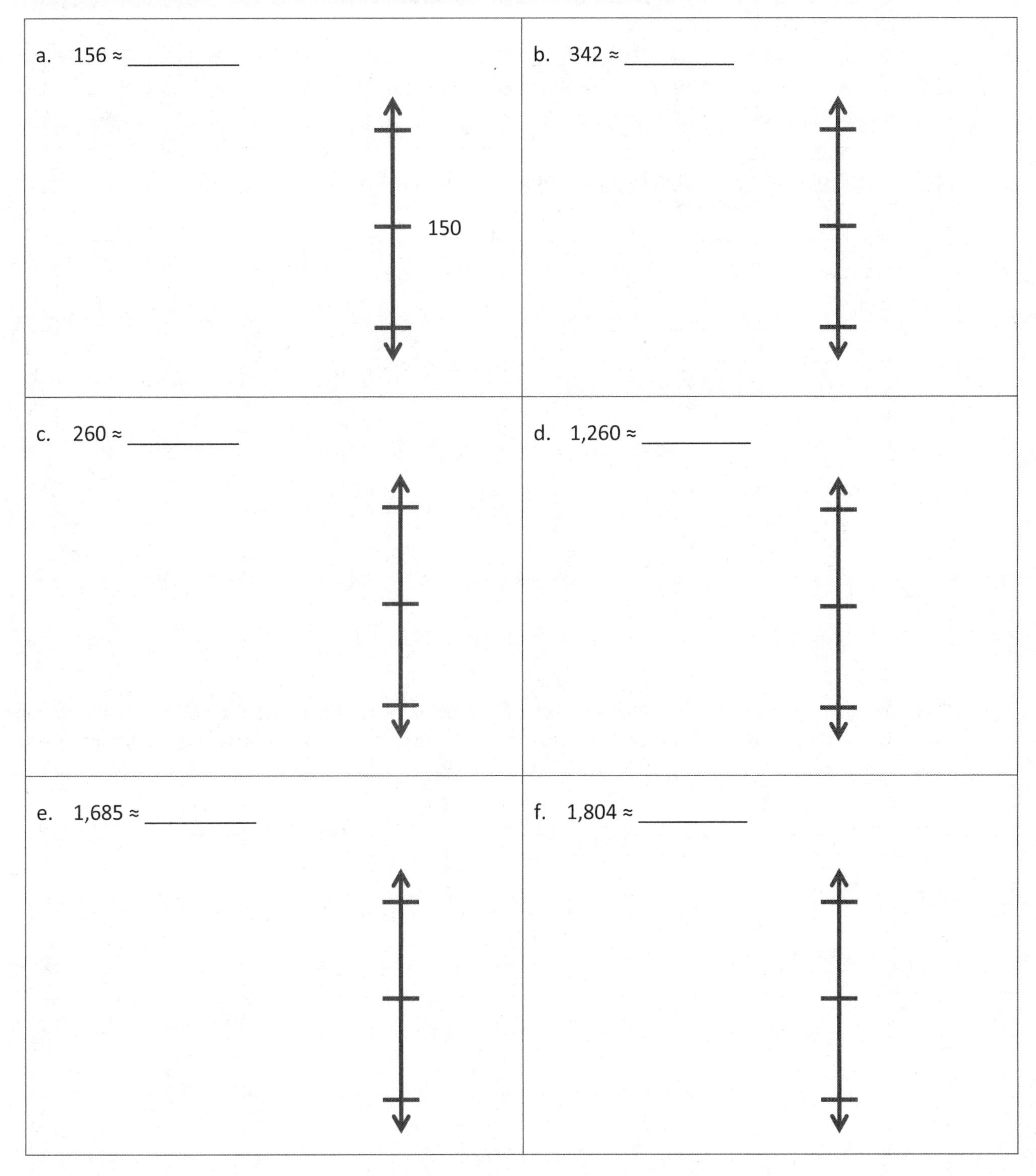

©2015 Great Minds. eureka-math.org
G3-M2-SE-B1-1.3.1-01.2016

2. Complete the chart.

a. Luis has 217 baseball cards. Round the number of cards Luis has to the nearest hundred.	
b. There were 462 people sitting in the audience. Round the number of people to the nearest hundred.	
c. A bottle of juice holds 386 milliliters. Round the capacity to the nearest 100 milliliters.	
d. A book weighs 727 grams. Round the weight to the nearest 100 grams.	
e. Joanie's parents spent $1,260 on two plane tickets. Round the total to the nearest $100.	

3. Circle the numbers that round to 400 when rounding to the nearest hundred.

368 342 420 492 449 464

4. There are 1,525 pages in a book. Julia and Kim round the number of pages to the nearest hundred. Julia says it is one thousand, five hundred. Kim says it is 15 hundreds. Who is correct? Explain your thinking.

©2015 Great Minds. eureka-math.org
G3-M2-SE-B1-1.3.1-01.2016

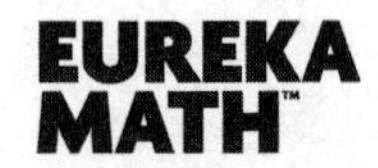

unlabeled place value chart

©2015 Great Minds. eureka-math.org
G3-M2-SE-B1-1.3.1-01.2016

Name ______________________________ Date ______________

1. Find the sums below. Choose mental math or the algorithm.

a. 46 mL + 5 mL

b. 46 mL + 25 mL

c. 46 mL + 125 mL

d. 59 cm + 30 cm

e. 509 cm + 83 cm

f. 597 cm + 30 cm

g. 29 g + 63 g

h. 345 g + 294 g

i. 480 g + 476 g

j. 1 L 245 mL + 2 L 412 mL

k. 2 kg 509 g + 3 kg 367 g

©2015 Great Minds. eureka-math.org
G3-M2-SE-B1-1.3.1-01.2016

2. Nadine and Jen buy a small bag of popcorn and a pretzel at the movie theater. The pretzel weighs 63 grams more than the popcorn. What is the weight of the pretzel?

? grams

44 grams

3. In math class, Jason and Andrea find the total liquid volume of water in their beakers. Jason says the total is 782 milliliters, but Andrea says it is 792 milliliters. The amount of water in each beaker can be found in the table to the right. Show whose calculation is correct. Explain the mistake of the other student.

Student	Liquid Volume
Jason	475 mL
Andrea	317 mL

4. It takes Greg 15 minutes to mow the front lawn. It takes him 17 more minutes to mow the back lawn than the front lawn. What is the total amount of time Greg spends mowing the lawns?

©2015 Great Minds. eureka-math.org
G3-M2-SE-B1-1.3.1-01.2016

Name ______________________________ Date ______________

1. Find the sums below. Choose mental math or the algorithm.

a. 75 cm + 7 cm

c. 362 mL + 229 mL

e. 451 mL + 339 mL

b. 39 kg + 56 kg

d. 283 g + 92 g

f. 149 L + 331 L

2. The liquid volume of five drinks is shown below.

Drink	Liquid Volume
Apple juice	125 mL
Milk	236 mL
Water	248 mL
Orange juice	174 mL
Fruit punch	208 mL

a. Jen drinks the apple juice and the water. How many milliliters does she drink in all?

Jen drinks ________ mL.

b. Kevin drinks the milk and the fruit punch. How many milliliters does he drink in all?

©2015 Great Minds. eureka-math.org
G3-M2-SE-B1-1.3.1-01.2016

3. There are 75 students in Grade 3. There are 44 more students in Grade 4 than in Grade 3. How many students are in Grade 4?

4. Mr. Green's sunflower grew 29 centimeters in one week. The next week it grew 5 centimeters more than the previous week. What is the total number of centimeters the sunflower grew in 2 weeks?

5. Kylie records the weights of 3 objects as shown below. Which 2 objects can she put on a pan balance to equal the weight of a 460 gram bag? Show how you know.

Paperback Book	Banana	Bar of Soap
343 grams	108 grams	117 grams

©2015 Great Minds. eureka-math.org
G3-M2-SE-B1-1.3.1-01.2016

Name ______________________________ Date ______________

1. Find the sums below.

a. 52 mL + 68 mL

b. 352 mL + 68 mL

c. 352 mL + 468 mL

d. 56 cm + 94 cm

e. 506 cm + 94 cm

f. 506 cm + 394 cm

g. 697 g + 138 g

h. 345 g + 597 g

i. 486 g + 497 g

j. 3 L 251 mL + 1 L 549 mL

k. 4 kg 384 g + 2 kg 467 g

©2015 Great Minds. eureka-math.org
G3-M2-SE-B1-1.3.1-01.2016

2. Lane makes sauerkraut. He weighs the amounts of cabbage and salt he uses. Draw and label a tape diagram to find the total weight of the cabbage and salt Lane uses.

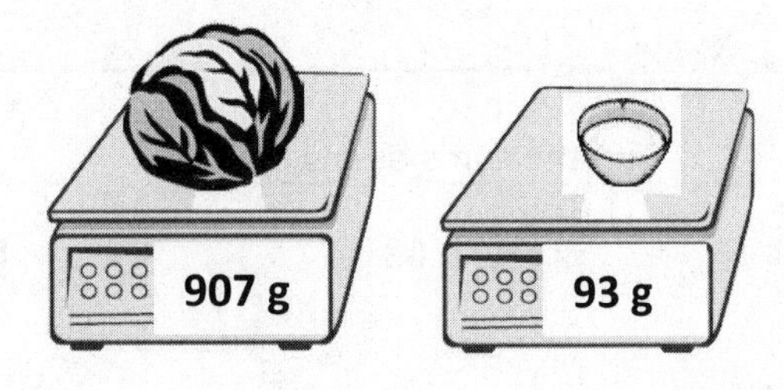

3. Sue bakes mini-muffins for the school bake sale. After wrapping 86 muffins, she still has 58 muffins left cooling on the table. How many muffins did she bake altogether?

4. The milk carton to the right holds 183 milliliters more liquid than the juice box. What is the total capacity of the juice box and milk carton?

©2015 Great Minds. eureka-math.org
G3-M2-SE-B1-1.3.1-01.2016

Name ______________________________ Date ________________

1. Find the sums below.

a. 47 m + 8 m

b. 47 m + 38 m

c. 147 m + 383 m

d. 63 mL + 9 mL

e. 463 mL + 79 mL

f. 463 mL + 179 mL

g. 368 kg + 263 kg

h. 508 kg + 293 kg

i. 103 kg + 799 kg

j. 4 L 342 mL + 2 L 214 mL

k. 3 kg 296 g + 5 kg 326 g

©2015 Great Minds. eureka-math.org
G3-M2-SE-B1-1.3.1-01.2016

2. Mrs. Haley roasts a turkey for 55 minutes. She checks it and decides to roast it for an additional 46 minutes. Use a tape diagram to find the total minutes Mrs. Haley roasts the turkey.

3. A miniature horse weighs 268 fewer kilograms than a Shetland pony. Use the table to find the weight of a Shetland pony.

Types of Horses	Weight in kg
Shetland pony	_____ kg
American Saddlebred	478 kg
Clydesdale horse	_____ kg
Miniature horse	56 kg

4. A Clydesdale horse weighs as much as a Shetland pony and an American Saddlebred horse combined. How much does a Clydesdale horse weigh?

©2015 Great Minds. eureka-math.org
G3-M2-SE-B1-1.3.1-01.2016

Name ______________________________ Date ______________

1. a. Find the actual sum either on paper or using mental math. Round each addend to the nearest hundred, and find the estimated sums.

A	B	C
451 + 253 = ____ ____ + ____ = ____	356 + 161 = ____ ____ + ____ = ____	652 + 158 = ____ ____ + ____ = ____
451 + 249 = ____ ____ + ____ = ____	356 + 148 = ____ ____ + ____ = ____	647 + 158 = ____ ____ + ____ = ____
448 + 249 = ____ ____ + ____ = ____	347 + 149 = ____ ____ + ____ = ____	647 + 146 = ____ ____ + ____ = ____
Circle the estimated sum that is the closest to its real sum.	Circle the estimated sum that is the closest to its real sum.	Circle the estimated sum that is the closest to its real sum.

b. Look at the sums that gave the most precise estimates. Explain below what they have in common. You might use a number line to support your explanation.

©2015 Great Minds. eureka-math.org
G3-M2-SE-B1-1.3.1-01.2016

2. Janet watched a movie that is 94 minutes long on Friday night. She watched a movie that is 151 minutes long on Saturday night.

   a. Decide how to round the minutes. Then, estimate the total minutes Janet watched movies on Friday and Saturday.

   b. How much time did Janet actually spend watching movies?

   c. Explain whether or not your estimated sum is close to the actual sum. Round in a different way, and see which estimate is closer.

3. Sadie, a bear at the zoo, weighs 182 kilograms. Her cub weighs 74 kilograms.

   a. Estimate the total weight of Sadie and her cub using whatever method you think best.

   b. What is the actual weight of Sadie and her cub? Model the problem with a tape diagram.

©2015 Great Minds. eureka-math.org
G3-M2-SE-B1-1.3.1-01.2016

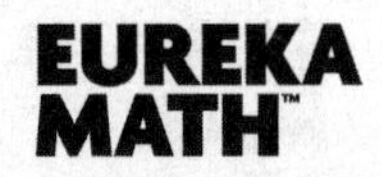

Name ______________________________ Date ________________

1. Cathy collects the following information about her dogs, Stella and Oliver.

Stella	
Time Spent Getting a Bath	*Weight*
36 minutes	32 kg

Oliver	
Time Spent Getting a Bath	*Weight*
25 minutes	7 kg

Use the information in the charts to answer the questions below.

a. Estimate the total weight of Stella and Oliver.

b. What is the actual total weight of Stella and Oliver?

c. Estimate the total amount of time Cathy spends giving her dogs a bath.

d. What is the actual total time Cathy spends giving her dogs a bath?

e. Explain how estimating helps you check the reasonableness of your answers.

©2015 Great Minds. eureka-math.org
G3-M2-SE-B1-1.3.1-01.2016

2. Dena reads for 361 minutes during Week 1 of her school's two-week long Read-A-Thon. She reads for 212 minutes during Week 2 of the Read-A-Thon.

   a. Estimate the total amount of time Dena reads during the Read-A-Thon by rounding.

   b. Estimate the total amount of time Dena reads during the Read-A-Thon by rounding in a different way.

   c. Calculate the actual number of minutes that Dena reads during the Read-A-Thon. Which method of rounding was more precise? Why?

©2015 Great Minds. eureka-math.org
G3-M2-SE-B1-1.3.1-01.2016

Name ______________________________ Date ______________

1. Solve the subtraction problems below.

a. 60 mL – 24 mL

b. 360 mL – 24 mL

c. 360 mL – 224 mL

d. 518 cm – 21 cm

e. 629 cm – 268 cm

f. 938 cm – 440 cm

g. 307 g – 130 g

h. 307 g – 234 g

i. 807 g – 732 g

j. 2 km 770 m – 1 km 455 m

k. 3 kg 924 g – 1 kg 893 g

©2015 Great Minds. eureka-math.org
G3-M2-SE-B1-1.3.1-01.2016

2. The total weight of 3 books is shown to the right. If 2 books weigh 233 grams, how much does the third book weigh? Use a tape diagram to model the problem.

3. The chart to the right shows the lengths of three movies.

The Lost Ship	117 minutes
Magical Forests	145 minutes
Champions	? minutes

   a. The movie *Champions* is 22 minutes shorter than *The Lost Ship*. How long is *Champions*?

   b. How much longer is *Magical Forests* than *Champions*?

4. The total length of a rope is 208 centimeters. Scott cuts it into 3 pieces. The first piece is 80 centimeters long. The second piece is 94 centimeters long. How long is the third piece of rope?

©2015 Great Minds. eureka-math.org
G3-M2-SE-B1-1.3.1-01.2016

Name ______________________________ Date ______________

1. Solve the subtraction problems below.

a. 280 g – 90 g

b. 450 g – 284 g

c. 423 cm – 136 cm

d. 567 cm – 246 cm

e. 900 g – 58 g

f. 900 g – 358 g

g. 4 L 710 mL – 2 L 690 mL

h. 8 L 830 mL – 4 L 378 mL

©2015 Great Minds. eureka-math.org
G3-M2-SE-B1-1.3.1-01.2016

2. The total weight of a giraffe and her calf is 904 kilograms. How much does the calf weigh? Use a tape diagram to model your thinking.

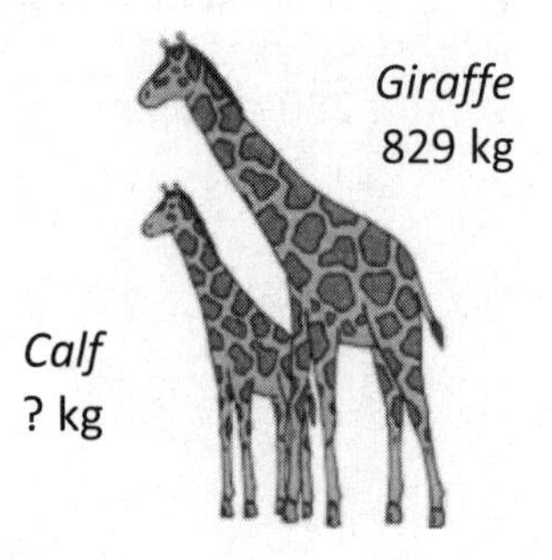

3. The Erie Canal runs 584 kilometers from Albany to Buffalo. Salvador travels on the canal from Albany. He must travel 396 kilometers more before he reaches Buffalo. How many kilometers has he traveled so far?

4. Mr. Nguyen fills two inflatable pools. The kiddie pool holds 185 liters of water. The larger pool holds 600 liters of water. How much more water does the larger pool hold than the kiddie pool?

©2015 Great Minds. eureka-math.org
G3-M2-SE-B1-1.3.1-01.2016

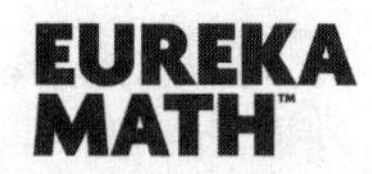

Name ______________________________ Date ________________

1. a. Find the actual differences either on paper or using mental math. Round each total and part to the nearest hundred and find the estimated differences.

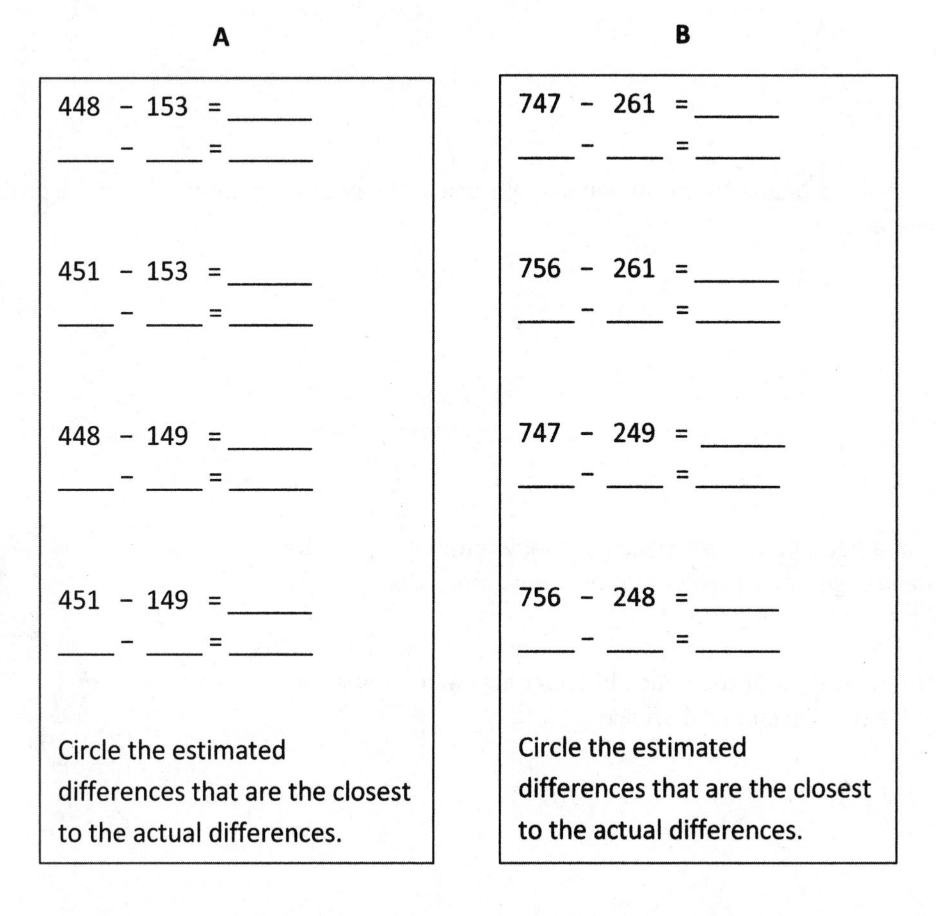

A
448 - 153 = _____
____ - ____ = _____
451 - 153 = _____
____ - ____ = _____
448 - 149 = _____
____ - ____ = _____
451 - 149 = _____
____ - ____ = _____
Circle the estimated differences that are the closest to the actual differences.
B
747 - 261 = _____
____ - ____ = _____
756 - 261 = _____
____ - ____ = _____
747 - 249 = _____
____ - ____ = _____
756 - 248 = _____
____ - ____ = _____
Circle the estimated differences that are the closest to the actual differences.

b. Look at the differences that gave the most precise estimates. Explain below what they have in common. You might use a number line to support your explanation.

©2015 Great Minds. eureka-math.org
G3-M2-SE-B1-1.3.1-01.2016

2. Camden uses a total of 372 liters of gas in two months. He uses 184 liters of gas in the first month. How many liters of gas does he use in the second month?

   a. Estimate the amount of gas Camden uses in the second month by rounding each number as you think best.

   b. How many liters of gas does Camden actually use in the second month? Model the problem with a tape diagram.

3. The weight of a pear, apple, and peach are shown to the right. The pear and apple together weigh 372 grams. How much does the peach weigh?

   a. Estimate the weight of the peach by rounding each number as you think best. Explain your choice.

   b. How much does the peach actually weigh? Model the problem with a tape diagram.

©2015 Great Minds. eureka-math.org
G3-M2-SE-B1-1.3.1-01.2016

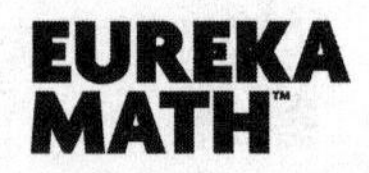

Name ______________________________ Date ______________

Estimate, and then solve each problem.

1. Melissa and her mom go on a road trip. They drive 87 kilometers before lunch. They drive 59 kilometers after lunch.

   a. Estimate how many more kilometers they drive before lunch than after lunch by rounding to the nearest 10 kilometers.

   b. Precisely how much farther do they drive before lunch than after lunch?

   c. Compare your estimate from (a) to your answer from (b). Is your answer reasonable? Write a sentence to explain your thinking.

2. Amy measures ribbon. She measures a total of 393 centimeters of ribbon and cuts it into two pieces. The first piece is 184 centimeters long. How long is the second piece of ribbon?

   a. Estimate the length of the second piece of ribbon by rounding in two different ways.

   b. Precisely how long is the second piece of ribbon? Explain why one estimate was closer.

©2015 Great Minds. eureka-math.org
G3-M2-SE-B1-1.3.1-01.2016

3. The weight of a chicken leg, steak, and ham are shown to the right. The chicken and the steak together weigh 341 grams. How much does the ham weigh?

   a. Estimate the weight of the ham by rounding.

   b. How much does the ham actually weigh?

4. Kate uses 506 liters of water each week to water plants. She uses 252 liters to water the plants in the greenhouse. How much water does she use for the other plants?

   a. Estimate how much water Kate uses for the other plants by rounding.

   b. Estimate how much water Kate uses for the other plants by rounding a different way.

   c. How much water does Kate actually use for the other plants? Which estimate was closer? Explain why.

©2015 Great Minds. eureka-math.org
G3-M2-SE-B1-1.3.1-01.2016

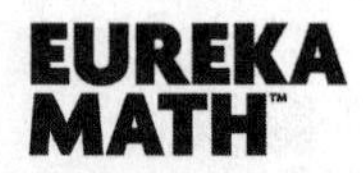

Name ______________________________________ Date ____________________

1. Weigh the bags of beans and rice on the scale. Then, write the weight on the scales below.

a. Estimate, and then find the total weight of the beans and rice.

Estimate: _______ + _______ ≈ _______ + _______ = ________

Actual: _______ + _______ = ________

b. Estimate, and then find the difference between the weight of the beans and rice.

Estimate: _______ – _______ ≈ _______ – _______ = ________

Actual: _______ – _______ = ________

c. Are your answers reasonable? Explain why.

©2015 Great Minds. eureka-math.org
G3-M2-SE-B1-1.3.1-01.2016

2. Measure the lengths of the three pieces of yarn.

Yarn A	_______ cm ≈ _______ cm
Yarn B	_______ cm ≈ _______ cm
Yarn C	_______ cm ≈ _______ cm

a. Estimate the total length of Yarn A and Yarn C. Then, find the actual total length.

b. Subtract to estimate the difference between the total length of Yarns A and C, and the length of Yarn B. Then, find the actual difference. Model the problem with a tape diagram.

3. Plot the amount of liquid in the three containers on the number lines below. Then, round to the nearest 10 milliliters.

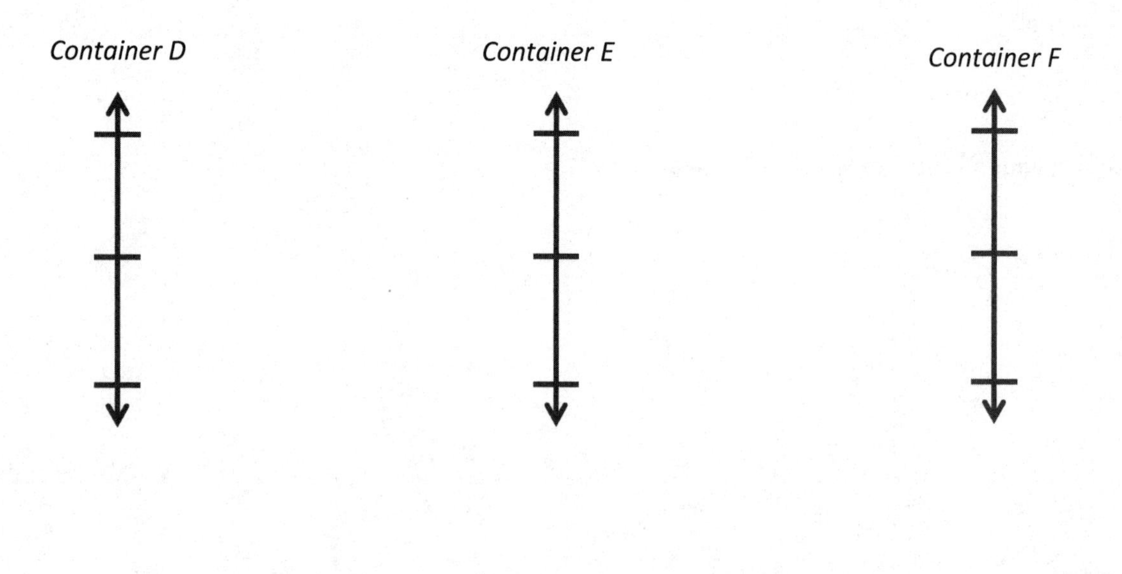

©2015 Great Minds. eureka-math.org
G3-M2-SE-B1-1.3.1-01.2016

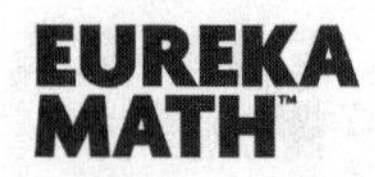

a. Estimate the total amount of liquid in three containers. Then, find the actual amount.

b. Estimate to find the difference between the amount of water in Containers D and E. Then, find the actual difference. Model the problem with a tape diagram.

4. Shane watches a movie in the theater that is 115 minutes long, including the trailers. The chart to the right shows the length in minutes of each trailer.

Trailer	Length in minutes
1	5 minutes
2	4 minutes
3	3 minutes
4	5 minutes
5	4 minutes
Total	

a. Find the total number of minutes for all 5 trailers.

b. Estimate to find the length of the movie without trailers. Then, find the actual length of the movie by calculating the difference between 115 minutes and the total minutes of trailers.

c. Is your answer reasonable? Explain why.

©2015 Great Minds. eureka-math.org
G3-M2-SE-B1-1.3.1-01.2016

Name ______________________________ Date ______________

1. There are 153 milliliters of juice in 1 carton. A three-pack of juice boxes contains a total of 459 milliliters.

   a. Estimate, and then find the actual total amount of juice in 1 carton and in a three-pack of juice boxes.

      153 mL + 459 mL ≈ ______ + ______ = ______

      153 mL + 459 mL = ______

   b. Estimate, and then find the actual difference between the amount in 1 carton and in a three-pack of juice boxes.

      459 mL − 153 mL ≈ ______ − ______ = ______

      459 mL − 153 mL = ______

   c. Are your answers reasonable? Why?

2. Mr. Williams owns a gas station. He sells 367 liters of gas in the morning, 300 liters of gas in the afternoon, and 219 liters of gas in the evening.

   a. Estimate, and then find the actual total amount of gas he sells in one day.

   b. Estimate, and then find the actual difference between the amount of gas Mr. Williams sells in the morning and the amount he sells in the evening.

©2015 Great Minds. eureka-math.org
G3-M2-SE-B1-1.3.1-01.2016

3. The Blue Team runs a relay. The chart shows the time, in minutes, that each team member spends running.

Blue Team	Time in Minutes
Jen	5 minutes
Kristin	7 minutes
Lester	6 minutes
Evy	8 minutes
Total	

   a. How many minutes does it take the Blue Team to run the relay?

   b. It takes the Red Team 37 minutes to run the relay. Estimate, and then find the actual difference in time between the two teams.

4. The lengths of three banners are shown to the right.

Banner A	437 cm
Banner B	457 cm
Banner C	332 cm

   a. Estimate, and then find the actual total length of Banner A and Banner C.

   b. Estimate, and then find the actual difference in length between Banner B and the combined length of Banner A and Banner C. Model the problem with a tape diagram.

©2015 Great Minds. eureka-math.org
G3-M2-SE-B1-1.3.1-01.2016